2004

HUMAN CLONING

Human Cloning

SCIENCE, ETHICS, AND PUBLIC POLICY

Edited by
Barbara MacKinnon

University of Illinois Press
URBANA AND CHICAGO

Library of Congress Cataloging-in-Publication Data
Human cloning : science, ethics, and public policy /
edited by Barbara MacKinnon.
p.   cm.
Includes bibliographical references (p.   ) and index.
ISBN 0-252-02491-5 (cloth : acid-free paper)
1. Human cloning—Moral and ethical aspects.
I. MacKinnon, Barbara, 1937–
QH442.2.H875   2000
174'.25—dc21   99-050519

C 5 4 3 2

Contents

HUMAN CLONING

BARBARA MACKINNON

Introduction

Suppose you have just been informed that you are a clone. After a moment of shock or disbelief, your first response might be a question. What does this mean? It could mean that you are one of a group of genetically identical individuals, the group itself more properly called the clone. Possibly you and the other members of the group were produced in a fertility clinic by splitting a fertilized embryo into a number of pieces, each being able at that point to become a separate embryo and individual. You were simply raised apart. You might be curious about your identical triplets or quadruplets, all the same age as you, similar to identical twins. Or it could mean that you were produced in a process like that of the famous sheep Dolly. Then you would have been produced from the body cell of another person and would be a younger identical twin of this person. There is no question in your mind that you, though a clone, are an individual. However, again, you might be concerned or curious about the person from whom you were cloned—who is also genetically identical to you—wanting to know the ways in which you are similar and different. Also, you might feel that there was something not quite right about the way in which you came to be, the motives behind it, possible harmful effects on you, and whether such manipulations are morally acceptable or ought to have been permitted.

We are now facing this possibility and can ask this same question, but in such a way that we use the answer to decide whether we want to or ought to go down that path. Until the birth of Dolly, it was generally believed that human cloning was either science fiction or a far-off future eventuality, not a real possibility. But now we know differently. Dolly changed things. She was cloned from the cells of an adult sheep and was born on July 5, 1996. Her birth was the result of efforts of two animal researchers, Ian Wilmut and Keith Campbell, who worked at the Roslin Institute in a small town seven miles from Edinburgh, Scotland. In attempting to produce sheep that would give insulin or other drugs in their milk, they drew upon some cells they had in the deep freezer, cells that had come from the mammary gland of a six-year-old ewe. Dolly was a clone, a genetic copy, of this ewe.

At once the event was hailed as revolutionary. Even though efforts to clone animals had been going on for more than forty years, many people had come to believe that producing a new animal from a cell that had taken on the specialized role of liver, skin, or mammary gland cell was not possible. Moreover, because the source of the genetic material was taken from an adult sheep, they could know in advance much about what the newly created lamb would be like. And Dolly was also a mammal, like us. At once we jumped to the conclusion that we would be able to clone human beings sooner than we expected and that this was no longer just science fiction.

Some found this prospect to be wonderful and exciting. However, the more prevalent initial reaction was one of fear and horror. Some people concluded that we could really make genetically identical copies of Hitler, as described in the 1976 book (and later movie) *The Boys from Brazil*, or drones to do menial labor as in some brave new world. Was this not a very contemporary example of the Frankenstein story, the possibility that science could create monster human beings that it could not control? Perhaps it was a scary possibility because, even more than other reproductive technologies, it showed the degree to which we could remake ourselves. We could reengineer our species. It also raised questions about our very humanness. It seemed to threaten the notion that every human being was a unique individual.

It did not take long for politicians and others to call for laws to ban the cloning of human beings. President Clinton had already established the National Bioethics Advisory Commission and so charged it with the new task of studying and advising him on human cloning. After a condensed period of review, in June 1997, the commission called for a moratorium on all efforts to clone a human being. It also called for a five-year period of study, after which the issue would be revisited.

In light of the call for public conversation on this matter, the College of Arts and Sciences at the University of San Francisco (USF) sponsored the "Human Cloning: Science, Ethics, and Public Policy" conference on April 3 and 4, 1998. This book is primarily a collection of revised essays from the conference. The conference focused on three areas—the scientific, ethical, and public policy aspects of human cloning—and this distinction is the basis for the threefold division of this book. Although these three areas overlap, the division is useful. It makes the reasonable assumption that in order to assess something, we should first know what it is. It also assumes that the realms of morality and public policy are distinct even if related in some ways. The conference wanted to consider the issue broadly and from a variety of perspectives. This collection includes views that are sympathetic to human cloning as well as some morally opposed to it.

The Science of Cloning

In order to know what to think about the cloning of human beings, we must first understand the science of cloning. What is cloning? How did it come about? What is now possible? For what purposes might cloning be useful? The first selection in this book, by George E. Seidel, an active animal cloning researcher, explains the meaning of the terms *clone* and *cloning* and describes the different types of processes now used to clone animals. It explains the difference between cloning produced by fission, or the cutting of embryos, and cloning produced by fusion or nuclear transfer. It is this latter process that produced Dolly, a process in which the nucleus of an unfertilized egg is sucked out and replaced by genetic material from the cell of a donor animal. Seidel details the way in which this is done and notes the success rates with various donor cells. He also points out some of the problems associated with the process, such as the number of tries needed to produce a healthy newborn (277 for Dolly) and the risks of abnormalities sometimes associated with the manipulations themselves.

It is interesting that some of this work is being done with farm animals, including sheep, cows, and pigs. Companies seeking commercial gain often fund the work, as was in the case of Dolly. For example, one aim is to produce better cows, ones that give better meat or more milk. Another aim is to make pharmaceutical products. For example, goats and cows that secrete certain drugs in their milk could enable us to produce and distribute these drugs much more efficiently and cheaply than by present means. Cloning could be used to produce transgenic animals, such as pigs or monkeys that contain some human genes. Their organs might be used for transplantation or as models in which to study human disease. Thus Polly the lamb was also produced at the Roslin Institute (born July 1997). She was cloned from sheep skin cells to which a human gene had been added. Cloning could be used as a means of basic research, designed, for example, to show how cells age, how they differentiate, how they sometimes multiply out of control, and how nerve cells might be rejuvenated.

The essay by Richard Lewontin also tells us about what cloning is as well as what it can and cannot do. For example, he shows why cloning cannot make identical copies of an animal or person any more than it could make identical copies of cloned plants. The environment, both in the uterus and in the egg as well as in development, also plays a role in what the mature plant or animal will become. Lewontin provides many interesting examples in his criticism of the general theory of genetic determinism. He shows that the organism does not compute itself from its genes.

In the essay by Philip Kitcher, some of the history behind the recent events that produced Dolly is outlined. To understand where cloning is today, it is useful to understand how it came about. As early as the turn of the twentieth century, early embryologists who were studying parthenogenesis, or producing organisms from eggs alone, attempted to produce identical twin embryos by splitting them.[1] However, it was not until the 1950s that Robert Briggs and Robert McKinnell, working on frog eggs, attempted cloning through nuclear transfer. They inserted a frog blastula cell into a frog egg, but couldn't get the tadpole cells to work. In the 1960s John Gurdon tried to insert into the frog egg what were thought to be specialized cells from the lining of tadpole intestines and was able to get the clone to the tadpole stage. In 1979 Karl Illmensee claimed to have cloned three mice. However, these claims were challenged and the truth was never quite satisfactorily determined. In 1987 Neal First and Steen Willadsen worked on cloning cows from embryos, using unfertilized cow eggs. They developed ways of manipulating embryos that led to present-day procedures.[2]

As one surveys this history even briefly, one can see that a great deal of effort by many people over a long period of time led to the production of Dolly. In fact, shortly before Dolly was born Wilmut and Campbell had produced two other lambs, Megan and Morag. These two lambs were cloned from embryo cells that had begun to differentiate. They had used a new process in which they starved the cells so that they could synchronize their cycles. This led them to believe that their process would allow them to use adult specialized somatic cells, as they did in producing Dolly. Moreover, because their aim was to add genes to the donor cells before implanting them in the enucleated egg, they hoped to be able to use adult cells because they worked better for this purpose. This raised the possibility of cloning adult human beings.

Cloning and Other Reproductive Technologies

As we contemplate the cloning of human beings, we can wonder whether there is anything really very new about this process. Is it not similar to other reproductive technologies in what it is designed to do? Is it not one more way to manipulate sperm and eggs to produce human beings in a laboratory rather than in the natural way? A child can now be produced from frozen donor eggs and sperm, gestated in a surrogate mother, and raised by people who are neither the genetic or gestational parents. Even Dolly had at least three "parents": the six-year-old ewe who provided the mammary cell, the sheep who was the source of the egg in which the ewe's cell was implanted, and a third sheep in whose uterus the growing embryo was placed and who gave birth to

Dolly. Sperm can now be injected into an egg when they would not otherwise be able to penetrate it. Eggs fertilized in the laboratory can be screened before implantation in a woman's uterus. Cloning thus could be seen as one more tool to help in the treatment of infertility. For example, cloning through fission might avoid the problem of multiple ovulatory cycles, for many embryos could be produced from one and then some could be frozen for future implantation. Cloning through nuclear transfer could help solve some of the problems of inheritance of genetic disease. The essays by John Robertson and Andrea Bonnicksen allude to some of these possibilities in their discussions of public policy. Those who find all of these reproductive technologies ethically problematic will also find cloning problematic. But should those who find these other technologies in themselves unproblematic also find cloning unproblematic? Is there any significant difference?

One thing that is interestingly different about cloning through nuclear transfer is that the immediate genetic source of the new individual is not two parents but one, the donor of the genetic material. There is a sense in which we might also say that there are two parents: the parents of the donor. Nevertheless, the child born through this type of cloning does not come immediately from the mixing of two lines of ancestors, as in naturally produced children. Other differences also exist, and some of these are discussed in the various essays in this collection.

Religious Perspectives

There is no specific group of readings in this collection devoted to religious perspectives and arguments regarding human cloning primarily because this was not included in the originating conference, which stressed philosophical rather than religious ethics. Still, in a number of places in this collection authors allude to or make use of religious arguments. For example, Jorge Garcia refers to the views of the Protestant theologian Gilbert Meilaender, and Alta Charo discusses certain deontological arguments that were similar to religious views presented before the President's National Bioethics Advisory Commission.

Many people depend on religion for their ethical beliefs. However, moral arguments can also be advanced that do not rely on religious sources. Thus we differentiate moral philosophy and religious ethics. Among religiously based arguments, one of the most difficult claims to substantiate is that God does not or would not approve of human cloning. One way to think about such claims is to consider a distinction made by Socrates in the Platonic dialogue *Euthyphro*. There Socrates implies that it is not because the gods approve of something that it is good, just, or holy. Rather, it is first just or holy and for this reason the gods approve or would approve of it.[3] The task would then be to examine cloning itself to see whether it seems right or wrong in itself.

One of the most often-used phrases by critics regarding human cloning is that it would be "playing God." Although this reason sounds religious, it is difficult to know exactly what is intended. Does it suggest a kind of hubris or pride in that we think we can take over the process and create ourselves? Is it that we do not know enough to know what we are doing and may really mess things up, for only God is so wise? If this is the issue, we would want to know why in this case we are not capable of judging well while in other cases we are. Or does the "playing God" criticism rely on a belief that human cloning is interfering with nature and that nature's way is God's way and that that way is best? If so, then we would need to examine these assertions, and in this we would rely on our reasoning process, as do the authors in this collection. In doing so, we leave the realm of religion and return to philosophy and philosophical analysis of issues surrounding human cloning. Some of these may relate to questions regarding the nature of the self and others to ethical matters.

Individuality and the Self

One of the reasons why many people may be so concerned about the cloning of humans is that it seems a threat to our identity, to our uniqueness as individuals. Instead of being a surprising new combination of the genetic contribution of two, the cloned child is somewhat of a repeat of its source. Today we try to see value in diversity and to appreciate differences. Each person is unique. There will never be another you (or ewe, as the pun goes). On the other hand, the images that accompany news accounts and discussions of cloning are images of sameness (multiple look-alike sheep, for example). Cloning makes us reflect on just what makes us similar to and different from one another.

Obviously, identical twins are different in many ways. However, multiple compact discs are almost identical. The CDs are made from the same material and also have the same form. Identical twins are produced through fission and are thus also made from the same stuff. However, they have environmentally caused and self-chosen differences in characteristics or form. Or consider two apples. They are individually different but still both apples. We are all human beings even though we are also different human beings. We share much in common with other humans and other members of our sex as well as with other vertebrates and all other living things.

For some, human cloning seems to pose a threat to the unique individuality of human beings. What is it that makes us who we are as particular selves? I know that it must feel different to be me than it does to be you. But if we look alike, have a similar body image, do we experience things in similar ways? If so, are we really distinct individual selves? These questions may well lie at the heart of why human cloning is so troubling to so many people. We are not sure how to think about what and who we are as individual selves or persons. Questions about what constitutes personal identity have puzzled philosophers for some time. Cloning provides an occasion for all of us to think carefully about such questions.

Ethical Issues in Human Cloning

The first response of many people when they were faced with the real possibility for cloning human beings that Dolly represented was that it was ethically reprehensible. However, now that we have had more time to think about it, the mood seems to have changed. When people moved from intuitive response to examining reasons for opposing human cloning, the reasons did not seem quite so convincing to some. Others find good reasons to continue to oppose cloning. What are some of these reasons that seem to make human cloning morally acceptable or morally wrong?

The essay by Bonnie Steinbock gives a general overview of the moral arguments against cloning, including the "playing God" argument, and concludes that none of these arguments is sufficient to show that it is immoral in itself or necessarily. She discusses concerns about risks, about unconsenting subjects, about effects on family relations, about psychological expectations for the child, and about the inexplicable repugnance that many feel about cloning humans. For example, she does not think that any bad effects there may be would be a result of the cloning itself, but rather of how we would use it. Furthermore, she points out that we take on the responsibility for changing nature all the time, as when we cure disease, and thus cannot easily use the don't-fool-with-nature argument. She finds that repugnance over something, whether cloning or other behavior, can sometimes be based on prejudice. Her essay presents a marked contrast with that by Jorge Garcia. In these two essays we see encapsulated the two major types of moral perspectives on human cloning. Garcia finds human cloning to be a threat to human dignity and to proper and normal spousal love. Concerning judgments based on feelings of repugnance, he questions the position that we must know why or have reasons for our negative moral reaction to certain practices. They can be wrong, he believes, even if we do not know exactly why. In his critique of a number of procloning arguments and positions he outlines a number of reasons why he believes that human cloning should not be accepted.

The essay by Philip Kitcher also addresses the moral questions through a description and analysis of three of the more reasonable scenarios or uses for human cloning. Kitcher also raises questions about social justice and whether we ought to spend our public money on the limited goods that human cloning could produce while ignoring more serious social problems and goods, including improvement of the quality of human life for the poor and disabled.

Human Cloning and Public Policy

Should it be illegal to clone a human being? Should all research that might lead to this result be legally banned? To adequately answer questions such as these requires that we be clear about what kinds of things should be legally regulated and why. It is generally conceded that not everything that is morally objectionable ought to be legally prohibited. Matters of law and matters of morality are not identical. However, morality often plays a role in what social policies we think we ought to adopt and sometimes what laws we ought to institute. The authors of this section of the book address this and other issues related to human cloning and public policy.

In her article, R. Alta Charo addresses the relationship between moral concerns and public policy. In particular, she explores the role of the value of liberty and the nature of a pluralistic society in establishing policy. Although people's motives may play a role in whether their actions are morally praiseworthy, she questions whether laws can always be governed by these considerations. As a member of the National Bioethics Advisory Commission (NBAC), she provides insights into the process that the NBAC used to develop its 1997 recommendations. She also considers whether human cloning ought to be governed by tort law or voluntary moratoria, or whether the government ought to take a more hands-off approach to this practice.

The essay by Andrea Bonnicksen, a political scientist, begins with her prediction that cloning is likely to develop incrementally through small steps and the emergence of a variety of types of procedures and techniques. She surveys the many recent legislative attempts to control or ban human cloning in the United States at both the federal and state levels as well as in international organizations and European countries. Rather than knee-jerk reactions and a crisis type of response to the possibility of the cloning of humans, and rather than prior legislative restraints on science, she suggests that we develop a broader and more studied and informed policy and principles and that we give some attention to the principles found in international policies. This would better fit the manner in which human cloning is most likely to develop. In particular, she raises some questions that could guide the clinical medical community as they and other groups contribute to the formulation of policy in this area.

In his contribution, John Robertson outlines some possibilities for the regulation of cloning in animal and human research as well as human cloning itself once it has become a real possibility and when it has become safe and effective. Thus he addresses such issues as who may clone and rear, whether people ought to be able to clone themselves or their parents, and whether single men and women should be permitted to produce children through cloning. In particular, he takes a close look at the issues of informed consent and psychological screening. He does this in the context of his concern about what he calls fundamental procreative or reproductive liberties. Others might want to consider what limits ought to be put on reproductive freedoms. Ought society to restrict such freedoms for the sake of protecting children from harm (for example, while cloning is still risky)? We do allow great liberties to parents prospectively and do not seek to restrict people from reproducing even when there is reason to believe that they will not be good parents. These matters are rightly open to question and serious consideration.

Susanne Huttner, a scientist and academic research advisor, addresses the issue of human cloning and public policy in terms of regulating technology in general. She points out the significant differences between regulating technology as a process and regulations directed at the outcome or product. Cloning, she suggests, is best thought of as one tool for producing certain products and benefits and argues that we ought to focus on these products instead of the means by which they are produced.

The Future of Cloning

Recent advances in human stem cell research have increased the pace at which we move toward the reality of cloning human beings. Embryonic stem cells are the inner mass of cells of the blastocyst and exist in an undifferentiated state for a short period of time from approximately five to seven days after fertilization. When these cells are placed in a culture dish, they are called embryonic stem cells.[4] By themselves, they do not seem able to develop into a fetus because they lack the other structure promoting part of the blastocyst that forms the placenta. The hope is that these stem cells or some modification of them could be transplanted into the proper sites of patients with diseased or nonfunctioning tissues, as in Parkinson's, kidney disease, or diabetes. Being pluripotent, able to become many different types of tissue cells, they would multiply and take on the jobs of the diseased or missing cells. However, because of problems of rejection, another route would use the somatic cell nuclear transfer technique of cloning. A person's own cell DNA could be placed in an enucleated egg and the process of embryonic development initiated. At the blastocyst stage, stem cells genetically identical to the patient could be recovered and used in the treatment of his or her disease. This type of cloning is now being called therapeutic human cloning and is being distinguished from reproductive human cloning. Although the purposes are distinct, what is learned from research in the therapeutic area could spill over to the reproductive, thus making it more likely that advances in the latter would be speeded up.

Objections to this type of cloning and to stem cell research have been raised (for example, whether this is actually experimenting on and using human embryos in ways that are ethically objectionable). In fact, since 1978 no federal funds have been able to be used for human embryo research, and since 1996 Congress has attached a rider to the National Institutes of Health (NIH) budget that would prohibit funding research that involves destruction of embryos. Producing embryos from which to derive stem cells would involve their destruction. No such restrictions have been placed on research using private funding, so a number of labs have been proceeding with such work. In early 1999, the legal counsel of the Department of Health and Human Services ruled that it is within the legal guidelines to fund human stem cell research if the cells are obtained from private funds. More recently, the NBAC voted against making such a distinction between producing and deriving the cells and simply using them. One suggestion is to use germline cells of fetuses that are already dead as a result of legal abortion. However, these cells may not be identical to stem cells derived from embryos, and deriving stem cells from this source is very difficult. As of this writing, federally funded human stem cell research is still awaiting the development of NIH guidelines and the final report of the NBAC.[5]

No one can predict how the science of cloning will develop. Some suspect that privately funded human fertility clinics are proceeding quietly in this area and will go public with this only when they have a healthy, viable clone. Efforts are also ongoing in the realm of animal husbandry and pharmacology, as there is a serious and growing commercial interest in using cloning in these areas. If the cloning of human beings becomes scientifically feasible, it may be very difficult to ban it. On the other hand, we can and do generally believe it appropriate to regulate scientific research and industrial applications in other areas and may well consider it appropriate to do so for human cloning also. For example, regulations regarding consent, protecting experimental subjects and children, and delineation of family responsibilities are just some matters possibly appropriate for regulation.

At the point of this writing, Dolly has had an offspring through normal means, and this lamb, named Bonnie, seems to be healthy. The laboratory that produced her sent a letter to the journal *Nature* reporting that there is some reason to believe that Dolly is nevertheless not quite normal and may be older than her years since birth. They assert that there are preliminary data that "Dolly's cells had slightly stunted telomeres," which are associated with aging.[6] A group in Japan claims to have cloned two calves, and two scientists at the University of Hawaii report cloning fifty mice, both using a method similar to that used by Ian Wilmut to create Dolly. Clones of mice and cows have been created using somatic cells from animal's ears and tails (i.e., from cells that are clearly bodily or somatic cells). Conferences on the commercial opportunities in cloning and transgenics are being held. Basic research continues. Although California, Michigan, and Rhode Island have banned the cloning of humans, no federal laws have been passed in the United States to prohibit it, and the public sentiment as of this writing seems more positive toward human cloning than it was two or three years ago. Of course, events could change things dramatically overnight, or the science may develop incrementally over another decade or two. Only time will tell.

NOTES

I would like to acknowledge the members of the USF faculty who served as the conference planning committee, without whose help in organizing and putting on the conference this collection would not have been possible: Raymond Dennehy (professor of philosophy), Carol Chihara (professor of biology), Theodore Jones (professor of chemistry), Susan Heidenreich (associate professor of psychology), and Thomas Cavanaugh (associate professor of philosophy).

1. Gina Kolata, *Clone: The Road to Dolly, and the Path Ahead* (New York: William Morrow, 1998), 59ff.

2. Ibid.

3. Plato, *Euthyphro,* 10a–11a.

4. Shirley J. Wright, "Human Embryonic Stem-Cell Research: Science and Ethics," *American Scientist* (July–Aug. 1999): 352.

5. *New York Times,* Apr. 9, 1999, p. A19.

6. *New York Times,* May 27, 1999, p. A24. This article also notes that there is some dispute about the significance and interpretation of the data. See also *New York Times,* June 29, 1999, p. A13.

PART 1

The Science of Cloning

1 Cloning Mammals: Methods, Applications, and Characteristics of Cloned Animals

Clone is derived from the Greek *klon,* meaning "twig." The Greeks knew that if one broke a twig from some species of trees and planted it, a copy of the tree would result. There are several definitions of *clone* as applied to animals. One definition is "to reproduce asexually"; if an organism is divided into two, that is not sexual reproduction. For example, one can bisect a mammalian embryo to form identical twins. Although the embryo is the result of a sexual process, making two from one is asexual reproduction, and a form of cloning. Another definition is "to make a genetic copy or set of copies of an organism." Some people define cloning very narrowly as "fusion or insertion of a diploid nucleus into an egg (oocyte)." *Clone* also may denote either a set of genetically identical organisms or an individual member of a set.

Degree of Identity of Clones

Genetic Differences

Identical twins, triplets, and so on, for which I have coined the term *identical multiplets* (Seidel 1983), are natural clones. Any of the methods used to manufacture clones will result in less identical sets of organisms than identical multiplets that result naturally. Clones are never identical. Although individuals in a set of clones may start with identical chromosomal DNA, many mutations will occur during development of an organism as complicated as a mammal. Moreover, the mitochondrial genes may differ, as described later in this chapter.

Environment

Of course, there are differences among individual members of a clone that result from environment, such as differences in birth weight resulting from gestation in uteri of large versus small females. In many cases, such environmentally caused differences are obvious. For example, behavioral traits and personality can be quite different among genetically identical individuals in all mammalian species. These are treated in more detail for humans in other essays in this book.

Cytoplasmic Differences

One reason that clones made by some methods are not identical is that whereas the chromosomal genetic component is identical, the cytoplasmic components are not. The clearest example is mitochondria, the small organelles responsible for most energy production in cells. Mitochondria have their own DNA (about 16,000 base pairs) that specifies some of the proteins in the mitochondria. Mitochondria differ genetically among female lines, and are inherited maternally as inclusions in the cytoplasm of the ovum.

Mitochondrial genes probably are responsible for some of the conformational differences (other than size) between hinnies and mules, which are sterile crosses of horses and donkeys. A hinny is mothered by a jenny and sired by a horse, whereas a mule has a mare for a mother and is sired by a jack. That mules have horse mitochondria and hinnies donkey mitochondria, in each case inherited exclusively from the mother, can be demonstrated easily by techniques of molecular biology.

Epigenetic Differences

The term *epigenetic* describes some of the chance outcomes of random motion of molecules that result in differences between genetically identical animals. Two clear examples are the migration of melanoblasts and random X-chromosome inactivation in females, both of which occur during embryonic development.

Hair color is created by melanin granules inserted into the growing hair by melanocytes that reside in hair follicles. Melanocytes originate in a specific region of the developing embryo known as the neural crest. The premelanocytes, called melanoblasts in the embryo, migrate from the neural crest, dividing as they move by ameboid movement, and invade hair follicles as they migrate. Some areas of the body of fetuses of some breeds of animals secrete factors that prevent successful invasion of hair follicles. This results in white spots or large areas of white. Also, the migratory process stops at a specific stage of fetal development and occurs independently on the left and right sides; sometimes the invasion stops short, resulting in stripes of white on the middle of the face of horses, for example. Also, in some animals the invasion front does not reach the extremities of the body, resulting in "stockings" of white on legs or white tails or heads.

This coloring of the developing embryo is epigenetic; that is, the migration pattern depends somewhat on chance. There is not a genetic instruction for each hair follicle to accept or reject melanoblasts. This is illustrated in figure 1.1, which shows two sets of identical twin bulls. One can readily see that though similar, the coat color patterns are not exactly identical within twin sets. As an aside, gray hair is caused by the death of melanocytes in the hair follicles. The hair still grows, but with no pigment added.

All female mammals have two X chromosomes in every cell, one inherited from the mother and one from the father. However, in any given cell, with a few exceptions, only one of the X chromosomes is genetically active. It will be either from the mother or the father. This is illustrated by a tricolor cat that has an orange and white father and a black and white mother. Pigment genes for the color of melanin granules in cats and many other mammals are located on the X chromosome. In any given cell in such a cat, one X chromosome, selected at random during development, is inactivated. Hence, orange color splotches on such a cat are directed genetically by the paternally inherited X chromosome and areas of black color by the maternally inherited X chromosome.

If genetically identical copies of this cat were made, each would have a different pattern of orange and black. Even though each cell has the same set of X chromosomes, the alleles are expressed differently in dif-

Figure 1.1. Two sets of manufactured identical twin Holstein bulls. Note that coat color patterns are not quite identical within each of the pairs of these genetically identical twins. (Courtesy of the author.)

ferent cells. Thus, genetically identical female mammals, entirely by chance inactivation of either the maternally or paternally inherited X chromosome, will never be identical except in the very special case in which there is identical information on both X chromosomes, as would occur only in highly inbred lines. Note that such lines would be either orange or black pigmented, never both. Random X inactivation affects thousands of genes on mammalian X chromosomes and accounts for the fact that naturally occurring female identical twins are less identical than male identical twins; the same would apply to sets of clones.

Technical Procedures for Manufacturing Clones

Nature has invested a great deal in preserving uniqueness of the individual while maintaining variation within each species. Even the process of development after sexual reproduction is completed is vulnerable to influences resulting in phenotypic differences. All methods of mammalian cloning, except perhaps those in popular fiction and tabloid journalism, begin with very few embryonic cells, often only one or two. Development must then follow through the four-cell stage, eight-cell stage and so on, exposed to all the chance aspects of development described earlier and in the essay by Richard Lewontin.

A number of technologies are used to manufacture clones. These include culture of cells in vitro, embryo transfer, micromanipulation, cell fusion, and, for some methods, cryobiology.

Embryo Transfer

Embryo transfer is nothing more than removing embryos from the reproductive tracts of females (donors) and transferring them to the reproductive tracts of other females (recipients) (Seidel 1981). Between recovery and transfer, the embryo can be manipulated. One form of cloning, bisection of an embryo to produce identical twins, is carried out during this interval, and it has been in limited use commercially in the cattle breeding industry for about fifteen years, resulting in tens of thousands of offspring.

It also is possible to remove sperm and ova from donor animals before fertilization and place these gametes together in a petri dish for fertilization, followed by early embryonic development in vitro before transfer to the female reproductive tract for gestation to term. This is how "test-tube" babies originate. Most cloning procedures are undertaken during this interval.

In cattle, both the recovery and transfer of embryos can be accomplished nonsurgically, which makes the use of this technology to amplify the reproductive capacity of individual cows fairly practical and at the same time raises few ethical issues. Cows normally have a single calf per year. By treating a cow with fertility drugs before insemination, it is possible to recover a number of embryos (six is the average number for one reproductive cycle, but sometimes twenty or more embryos are produced) that can be transferred to reproductive tracts of other cows for gestation. Therefore, a cow of particular genetic or commercial value can be the genetic mother of a litter of calves, but at the expense of the genetic reproduction of the less valuable recipient cows.

In figure 1.2 is a photograph of a Holstein dairy cow, who was the genetic mother of ten Holstein calves, pictured with their birth mothers, ten crossbred beef cows. In taking this photograph, we observed a good lesson. The beef cows that had given birth to and nursed the calves were unaware that they were not the genetic mothers, and became very unhappy when they were separated from the calves while they were grouped for the picture. The noise of their bawling was deafening. The genetic mother showed no interest in the calves. What is important is who takes care of the kids, not where the genes came from.

Micromanipulation

Microsurgery on embryos requires variations on tools such as the lever and the screw. Micromanipulators and the various accessory

Figure 1.2. Holstein dairy cow with her ten genetic offspring and their surrogate mothers who gestated and nursed them. (Reprinted with permission from the cover of *Science* 211 [1981], © 1981 by the American Association for the Advancement of Science.)

microtools reduce the gross movements of the hand to small and precise movements best visualized under the microscope (fig. 1.3). The microtools (pipettes, blades, needles, and probes) are manufactured as needed in the laboratory, using special equipment such as a microforge.

Brief Historical Perspective

First the Newt

The first attempt to clone a vertebrate was undertaken by Spemann (1967 [1938]). Using a baby's hair as a tool, Spemann ligated a one-cell newt embryo to produce a dumbbell shape, so that one half contained only cytoplasm, and the other half the nucleus. When about sixteen nuclei had been replicated in the one half, Spemann loosened the hair ligature, allowing one nucleus into the half containing only cytoplasm before he again tightened the hair, dividing the embryo into two. One embryo was about four cell divisions behind the other. These identical twin newts were allowed to develop to form millions of cells each.

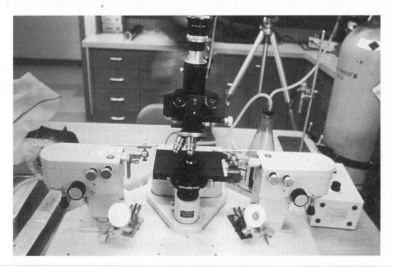

Figure 1.3. A microscope and a pair of micromanipulators. (Courtesy of the author.)

It is not known whether adult newts ever developed from such embryos, but the scientific point was illustrated.

Amphibians

There was a flurry of experimental activity in the 1950s and 1960s on cloning amphibia, often with embryonic donor cells (Briggs and King 1952), but also with adult somatic cells. Transfer of a nucleus from a somatic cell (generally the epithelium of the gut) of a tadpole to an ovum from which the original nucleus had been removed often resulted in an adult frog. When the nucleus from the somatic cell of an adult frog was transferred to an ovum, many of these transfers resulted in a tadpole, but development rarely continued to an adult frog. It was concluded that some characteristics of nuclei from adult cells mitigated against development to viable adult organisms when such nuclei were placed in the environment of an activated, enucleated ovum.

Methods of Cloning Mammals

Description of the Mammalian Ovum

A description of the mammalian ovum will provide perspective on some of the problems encountered in cloning as well as on some of the techniques that, seemingly against all odds, succeed. The unfertilized ovum (oocyte) of most mammals is a sphere of about 140 microns (micrometers), or 1/200 of an inch in diameter. Differences between oocytes from elephants, human beings, rabbits, and zebras are subtle matters of color, texture, and extracellular investments. The exception is the slightly smaller size of oocytes from rodents.

The oocyte is the largest cell in the body. If one holds a test tube containing an oocyte up to the light at just the right angle, the oocyte can barely be visualized with the unaided eye. It looks like a speck of dust. Oocytes are surrounded by an acellular coat called the zona pellucida, which remains around the embryo after fertilization until there are hundreds of cells (fig. 1.4). This thin, spherical coat is gelatin-like

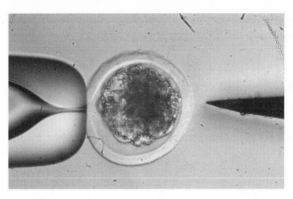

Figure 1.4. A fragment of a broken razor blade glued to a glass rod is being used to bisect this approximately eighty-cell bovine embryo. The embryo is immobilized by suction on the acellular coat, the zona pellucida. (Courtesy of the author.)

with a slightly brittle consistency, but it is considerably more flexible than a fingernail. The zona pellucida can be cracked open to remove the oocyte without the zona's losing shape; thus, the oocyte can be reinserted into the zona pellucida, which provides protection and supports the roughly spherical shape of the early embryo.

As the fertilized ovum develops by divisions of the component cells up to the sixteen-cell stage, the spherical cells touch each other tangentially. The total size and weight of the embryo does not change until there are about 100 cells, so cells keep getting smaller with each cell division up to this stage. At the sixteen-cell stage, however, the embryonic structure "compacts" so that cells, which earlier had touched other cells only at one point because of being perfectly spherical, now form contacts with adjacent cells over their entire surfaces and become polygonal. From this point on, cells cease autonomous development and begin to work in concert to form the organism. The cells inside the compacted embryo are a little different from those on the outside, and divide more slowly, so cells no longer divide synchronously after about the thirty-two-cell stage. If the zona pellucida is removed before compaction, the embryo dissociates into isolated cells that die after a few divisions; after compaction, however, the embryo maintains its shape, even outside the zona pellucida. Compaction is a significant event from the standpoint of several methods of cloning.

Method 1: Blastomere Separation

A blastomere is the name given to the cells that make up the early embryo. Clones of two, three, or sometimes four identicals can be produced by dividing the two- to eight-cell embryo into two, three, or four groups of blastomeres. The technique is built on one of the earliest experiments with mammalian cloning, which was carried out in Germany (Seidel 1952): One cell of a two-cell embryo was killed by macerating it with a needle and transferring the embryo to the reproductive tract of a recipient to see whether it would continue development. Normal offspring resulted from the remaining embryo cells in some cases.

About twenty years ago, one of my graduate students removed the zona pellucida from two-cell mouse embryos and cultured each half in a separate container. Some of the half-embryos developed apparently normally, but he never got both halves to develop to live offspring after embryo transfer. We considered two possibilities: Only one of the cells of the two-cell embryo actually contained the right material to make an individual (i.e., both cells were not totipotent), or our culture methods simply were not good enough to support a developmental rate such that both halves would develop to normal young with reasonable frequency. It turned out to be the latter problem. We then started working with bovine embryos and went back to nature to solve the problem, as did others, using the reproductive tract of the rabbit as an in vivo incubator by transferring embryos to the ligated fallopian tubes for an interval of several days. Recovered embryos were retransferred to the reproductive tract of a cow and resulted in normal calves.

The question of totipotency of individual blastomeres was answered eloquently by Willadsen (1979) in the United Kingdom. He placed each blastomere of a four-cell sheep embryo into a separate surrogate zona pellucida. The zonae were then embedded in a block of agar and transferred to the fallopian tube of an intermediate recipient (a sheep, in this instance). Each of these cells (one-fourth of an embryo) divided into two (two-eighths of an embryo), then again into four cells (four-sixteenths of an embryo), and, after five days, into little postcompaction embryos. At this stage, the embryos were recovered from the intermediate recipient, dissected out of the agar block, and retransferred to the reproductive tract of another sheep for gestation to term. Identical quadruplets resulted in some cases, demonstrating that each cell of a four-cell embryo was capable of forming a complete individual.

Nature is conservative. There are more cells in an embryo than are needed; indeed, as the embryo develops, the redundant cells die off. One-fourth the normal number of cells seems, nevertheless, to be the limit; only rarely do all four quarter embryos develop to term, so one often gets identical twins or triplets from such a manipulation. Further divisions to one-eighth embryos almost never produce viable fetuses. Willadsen repeated this experiment with horses, swine, and cattle, making identical twins and triplets.

Method 2: Splitting Embryos

Our laboratory developed a method of producing identical twins (Seidel 1983) that is akin to blastomere separation: splitting embryos. Using a pipette to apply gentle suction to hold the embryo in place and a fragment of a razor blade glued to a glass rod, one can bisect a postcompaction embryo, transfer one half to a surrogate zona pellucida, and return the embryos to the reproductive tract of an appropriate female for gestation (fig. 1.4). Identical twins result about 25 percent of the time (fig. 1.1), single births about 50 percent of the time, and no development the remaining 25 percent. It also is possible to divide an embryo in thirds, but the success rate is substantially lower.

This technique has been used commercially by the cattle industry for more than a decade because it is practical under field conditions and amplifies the number of offspring that can be obtained from valuable animals. The pregnancy rate following transfer of whole embryos is about 65 percent and that of bisected embryos is 50 percent per half, but there are two halves. Thus, the yield is about 50 percent more calves following bisection. This amplification of reproduction is the main purpose of splitting embryos in a commercial setting, not the fact that sets of genetically identical animals are produced. I estimate that nearly 50,000 calves have been produced worldwide using this technology.

From the standpoint of husbandry, there are a number of other applications of embryo splitting and transfer. For example, a breeder who uses this technology to produce identical twins could sell the best genetics and still retain them in his or her own breeding program by selling one twin and keeping one twin. Sometimes, the genetic value of a breeding animal can be evaluated only by assessing carcass qualities, which obviates the prospects of using a proven valuable animal for breeding. If one manufactures twin offspring, one animal could be slaughtered for evaluation and its twin used for breeding. Bisection of embryos can also be combined with freezing, such that it is possible to keep a "spare" on hand for later transfer should it become desirable to have an additional animal of a particular genetic makeup. In certain experiments studying maternal effects on reproductive traits, one-half of a female embryo can be frozen and one-half transferred and nurtured to adulthood, at which time the frozen half-embryo is transferred to the reproductive tract of its adult twin. Thus, an animal could give birth to her identical twin.

In a research setting, splitting is used to manufacture animals for experiments; with identical twin sets, fewer animals are required to obtain reliable information. In some cases, only one-third as many animals are needed to test hypotheses compared to not using identical twins; the power of this approach depends on the relative genetic contribution to the trait being studied. For example, assignment of one member of each of four sets of twins to a treatment group and the other member of each set to the control group can be as powerful an experiment as randomly assigning twenty unrelated animals to treatment and control groups. This use of fewer experimental animals is a strong ethical argument for the use of such biotechnology.

Method 3: Nuclear Transplantation or Cell Fusion

There are two advantages of using nuclear transplantation or cell fusion for cloning. First, although success rates are lower than with the methods just described, one is not limited to sets of four identical multiplets. Second, one is not limited to using embryonic cells as genetic donors.

All mammals are diploid genetically, and receive a haploid set of genes from each parent at fertilization. The goal of this approach to cloning is to use a diploid set of genes to program development into an individual while bypassing the step of fertilization. However, an oocyte is required to start the process of embryonic development. The important considerations in bypassing fertilization are that a diploid set of genes is already in a donor nucleus (which has a haploid set of genes from the father plus a haploid set from the mother), more than a complete diploid set of genes is detrimental, and the act of introducing the half-set from the father at fertilization normally activates the subsequent developmental process of the oocyte. A complete set of genes theoretically could be obtained by removing the nucleus of an embryonic cell any time after the first cell division, from a gestating fetus, or from an individual after birth. Until recently, success was obtained only from donor nuclei from certain cells of embryos up to the 100- to 500-cell stage (Willadsen 1986).

An unfertilized ovum recovered after ovulation is the required environment for the diploid nucleus of a cell to initiate embryonic development; however, such an ovum already has a haploid set of genes, and adding a diploid set would result in an abnormal triploid embryo that also would not be a clone. Moreover, because the ovum normally is activated by the act of sperm penetration, this activation now must be done artificially. In practice, the most common procedure for removing the haploid set of chromosomes from the oocyte is to aspirate them via a micropipet, although other procedures have been used, such as destroying them with a laser. After the chromosomes are removed, the cell containing the nucleus is fused with the oocyte using an electrical pulse (fig. 1.5). Thus one effectively is fertilizing the large oocyte with a small cell containing a diploid nucleus rather than a haploid sperm. The electrical pulse also substitutes for the activating role of the sperm. A variation on this process is to inject the donor cell into the cytoplasm of the oocyte instead of using electrical fusion; the oocyte still must be activated in some way.

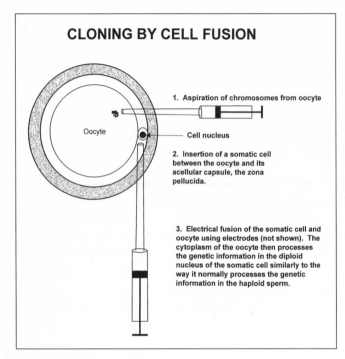

Figure 1.5. Cloning by cell fusion. (Courtesy of the author.)

The whole process consists of many steps. Oocytes are cultured for twenty-four hours to mature them, the chromosomes are removed from the oocyte, blastomeres are harvested from the donor embryo, cells are fused together, the resulting embryo is cultured to the early blastocyst stage of development, some of the cells are cryopreserved for future cloning, and the embryo is transferred to a recipient for gestation to term.

Success rates of the process are much less than 100 percent. For example, when starting with a sixteen-cell embryo and fusing each cell to sixteen one-cell oocytes, perhaps four, or 25 percent, would develop normally. However, with serial cloning, the next round would begin with 4 sixteen-cell embryos. If these were recloned with a 25 percent success rate, successive rounds would result in 16 sixteen-cell, then 64,

256, etc. genetically identical embryos. At any stage, some of those embryos could be frozen and others transferred to recipients. More than a thousand calves have been made using such procedures. The largest number of identical animals in a clutch that I am aware of is eleven. Note that success rates seem to diminish in some species after two or three rounds of serial nuclear transplantations for unexplained reasons.

Characteristics of Animals Cloned by Nuclear Transplantation or Cell Fusion

One of the most interesting findings from this research is that a number of the calves and lambs produced with these procedures were not quite normal. Some were too large to be born normally and had to be taken by cesarean section. A normal calf weighs about 90 pounds, but some of these weighed more than 150 pounds. Something was not quite right. The earlier methods of cloning I described, separating blastomeres and dividing embryos in half, result in normal offspring.

The logistics of the current technology of nuclear transplantation or fusion of cells require that, for cattle, the manipulated embryo develop in vitro for eight or nine days, so embryos develop sufficiently that nonsurgical methods can be used for embryo transfer. We hypothesize that some aspect of in vitro culture conditions perturbed some step in embryonic development. Although most offspring cloned by these procedures are morphologically normal, except that about 20 percent have increased size, many seem slow to nurse and are metabolically abnormal (e.g., they have low blood glucose levels at birth). Thus, some cloned calves and lambs die at birth unless special nursing care is provided. What is remarkable about these abnormal calves and lambs is that they become entirely normal if they survive the first few days. They develop into normal adults and produce normal calves. Thus, whatever causes enlarged or otherwise abnormal calves is not genetic in origin; it is not transmitted to the next generation. Our best explanation is that the placenta is abnormal.

Success rates of cloning by cell fusion are low and variable, placentas are abnormal, and some calves and lambs are epigenetically abnormal. With serial cloning, the percentage of epigenetically abnormal offspring increases. Perhaps these problems are less pronounced in species other than cattle and sheep. Also, it appears that such problems are being circumvented by improved procedures of in vitro culture of embryos.

The birth of the famous sheep Dolly demonstrated that it is possible to obtain the nucleus for successful transplantation from the cell of an adult, in this case mammary tissue (Wilmut et al. 1997). Despite very low success rates (Dolly represented one success in 277 tries), this was a significant scientific event. Ironically, development of techniques to clone adult cells was a byproduct of research to produce transgenic animals that produce drugs in milk for the pharmaceutical industry. Many of the lambs or fetuses generated during the Dolly experiments had the same kinds of abnormalities as described for giant calves; in some cases the abnormalities were so severe that the pregnancy aborted.

As of this writing, the world's production of cloned animals probably includes fewer than 1,000 animals produced by blastomere separation, about 50,000 twins and triplets (of which 98 percent are cattle) produced by splitting embryos, nearly 2,000 animals (of which about 1,800 are cattle) produced by cell fusion techniques using embryonic cells as nucleus donors, and perhaps fifty lambs and calves resulting from fetal fibroblasts as donors. Dolly represents the first of a new group of clones derived from nuclei of adult cells. Currently, this includes dozens of calves and hundreds of mice derived from nuclei of cells from many different tissues.

Cloning Procedures: A Technological Continuum

Cloning methodology represents a technological continuum. Some would argue that dividing embryos in half is not cloning, but rather only making identical twins. Actually, identical twins are very valuable scientific subjects. Naturally occurring identical twins of several species have been used for experiments for many years. A variation of this concept is producing highly inbred lines resulting in animals that are, for practical purposes, identical with each other. Crossing inbred lines produces outbred animals still genetically identical with each other, except, of course, that the males and females differ.

The next step in the continuum was to produce identical multiplets by separating blastomeres or splitting embryos instead of relying on naturally occurring identicals or using inbred lines. Fusion of blastomeres with oocytes is a further step in the continuum; serial cloning via blastomere fusion, making as many copies as desired, represents a further step. The next advance was obtaining donor nuclei from embryonic somatic cells, the fetal fibroblasts mentioned earlier. This was the most surprising step in the continuum for most scientists working in this area. It was described in a paper from the same group that made Dolly, but published a year earlier (Campbell et al. 1996). Much of the world did not notice this important publication. Use of newborn somatic cells for donor nuclei is the next logical step and is progressing with some success. Thus, there is a continuum of technologies. Based on previous findings, scientists working in the area were not all that surprised when Dolly came along.

Applications of Cloning Technology

There are numerous potential applications of cloning. From an agricultural standpoint, an obvious application is obtaining two for one; more offspring are produced by dividing embryos in half than not. We can also obtain certain basic information and improve experiments by producing genetically identical individuals. The chance of pregnancy in an individual female increases when two half embryos are transferred rather than one whole embryo. This is an argument for dividing human embryos in half for in vitro fertilization programs for infertile women. There may be a situation in which only one embryo is available, so should it be divided in half to increase the chances of pregnancy? This is a very difficult question. This procedure has not been done with human embryos, and success rates might be very low because of species differences.

An important application of cloning is to produce transgenic animals to produce valuable pharmaceuticals, for example in their milk. The big advantage is that genetic changes can be made in somatic cells in vitro, and when the proper change is made, an animal is produced with a nucleus from such a genetically changed cell.

There are numerous potential applications. Nuclear transplantation or cell fusion could be used to circumvent certain mitochondrial diseases. We could think of a woman who produces an embryo that will form an abnormal baby because the cytoplasm of her cells contains abnormal mitochondria. However, nuclei from the resulting embryonic cells could be placed into the cytoplasm of an ovum of another woman who has normal mitochondria. So in this case, nuclear transplantation is just a method of circumventing the problem; technically an embryo is being cloned. Another application is to copy an outstanding adult animal when one didn't have the foresight to freeze genetically identical cells when they were embryos. One might want to replace a favorite pet. I get a lot of phone calls about this. This is nonsense for the same reason that one would not get an identical human when cloning, and certainly not the same person. Similarly, a clone will not be the same pet. The pet may look like the initial individual, for example in color, but will be an entirely new individual, with a different personality from the nucleus donor. One could reproduce a prechosen genotype, and from an agricultural standpoint that makes some sense. A particular cow, for example, may be genetically adapted to produce large quantities of milk in the tropics, and we might want to copy that individual via cloning. Another sound application would be to make a genetic copy of a valuable bull for breeding purposes.

Probably one of the most important applications of human cloning research will eventually be to copy tissues for transplantation. It is likely that embryonic cells can be made to develop into pancreas or liver tissue, for example, without going into the stage of producing a fetus with a central nervous system and so on. Such tissues would not be rejected by the immune system if donor cells originated from the person receiving the tissue. This technique might even be used to produce nerve cells for people with spinal injuries.

Ethics of Cloning Animals

I will briefly consider ethics of cloning from the animal standpoint. The ethics of cloning people are covered elsewhere in this volume. When working with agricultural and experimental animals, personnel should be trained appropriately. Experimental designs should be sound. Experiments should be done for good reasons and be well thought out. Animals should receive excellent care, and any pain and suffering should be minimized (for example, by using effective anesthesia and analgesics). If an animal is abnormal at birth and appears to be suffering, it should be euthanized quickly. Procedures to make those types of decisions should be in place. Some attention must be devoted to preserving gene pools for maintaining genetic variation when we do cloning experiments. It could be risky to have only a few genotypes of pigs or cattle if some unanticipated pathogen evolves or the environment changes.

Summary

Scientists studying fertilization and embryology developed methods of cloning in the course of their research. When cloning became an objective, this further increased our knowledge of embryology, and this symbiotic process continues in an expanding spiral of increasing knowledge. The information and expertise from such work can be used in a variety of ways. Although sinister applications can be conjured up, so can wonderful outcomes such as providing replacement tissues for people otherwise condemned to die prematurely, providing a child to a childless couple, or copying a prize bull. Although we clearly must be careful so our efforts are not used inappropriately, it seems to me that our major responsibility is to put cloning technology to beneficial use. A final point is that every individual, cloned or not, is unique. Cloning is not a means to live forever.

REFERENCES CITED

Briggs, R., and T. J. King. 1952. "Transplantation of Living Nuclei from Blastula Cells into Enucleated Frogs' Eggs." *Proceedings of the National Academy of Sciences, U.S.A.* 38:455–63.

Campbell, K. H. S., J. McWhir, W. A. Ritchie, and I. Wilmut. 1996. "Sheep Cloned by Nuclear Transfer from a Cultured Cell Line." *Nature* 380:64–66.

Seidel, F. 1952. "Die Entwicklungspotenzen Einer Isolierten Blastomere des Zweizelstadiums im Säugertierei [Developmental Potential of an Isolated Mammalian 2-Cell-Stage Blastomere]." *Naturwissenschaften* 39:355–56.

Seidel, G. E., Jr. 1981. "Superovulation and Embryo Transfer in Cattle." *Science* 211:351–58.

————. 1983. "Production of Genetically Identical Sets of Mammals: Cloning?" *Journal of Experimental Zoology* 228:347–54.

Spemann, H. 1967 [1938]. *Embryonic Development and Induction.* Rpt., New York: Hafner.

Willadsen, S. M. 1979. "A Method for Culture of Micromanipulated Sheep Embryos and Its Use to Produce Monozygotic Twins." *Nature* 277:298–300.

————. 1986. "Nuclear Transplantation in Sheep." *Nature* 320:63–65.

Wilmut, I., A. E. Schnieke, J. McWhir, A. J. Kind, and K. H. S. Campbell. 1997. "Viable Offspring Derived from Fetal and Adult Mammalian Cells." *Nature* 385:810–13.

2 Cloning and the Fallacy of Biological Determinism

Much of the motivation for human cloning and many of the ethical dilemmas that are said to be raised by cloning rest on the mistaken synecdoche that substitutes "gene" for "person." Why is there any motivation for cloning a human genome? In one scenario a self-infatuated parent wants to reproduce his perfection or a single woman wants to exclude any other contribution to her offspring. In another, morally more appealing story, a family is in an accident that kills the father and leaves an only child on the point of death. The mother, wanting to have a child who is the biological offspring of her dead husband, uses cells from the dying infant to clone a baby. Or what about the sterile man whose entire family was exterminated in Auschwitz and who wants to prevent the extinction of his genetic patrimony? Creating variants of these scenarios is a philosopher's parlor game. All such stories appeal to the same impetus that drives adopted children to search for their "real" (i.e., biological) parents in order to discover their own "real" identity. They are modern continuations of an earlier preoccupation with blood as the carrier of an individual's essence and the mark of legitimacy. It is not the possibility of producing a human being with a copy of someone else's genes that has created the difficulty or that adds a unique element to it. It is the fetishism of the blood, which, once accepted, generates an immense array of apparent moral and ethical problems. If not for the belief in blood as essence, much of the motivation for the cloning of humans would disappear. In like manner, the ethical problems raised by the question of individuality would disappear if it turned out that cloning a set of genes did not reproduce the "person" from whom they were taken, but simply another, different individual, unique but with much the same nose shape as the genetic donor. But the genetic determinism on which these problems rest is an error of understanding of how organisms develop.

The fallacy of genetic determinism is to suppose that the genes make the organism. It is a basic principle of developmental biology that organisms undergo a continuous development from conception to death, a development that is the unique consequence of the interaction of the genes in their cells, the temporal sequence of environments through which the organisms pass, and random cellular processes that determine the life, death, and transformations of cells. As a result, even the fingerprints of identical twins are not identical. Their temperaments, mental processes, abilities, life choices, disease histories, and deaths certainly differ despite the determined efforts of many parents to enforce as great a similarity as possible. Twins often are given names with the same initial letter, are dressed identically with identical hair arrangements, and given the same books, toys, and training. There are twin conventions at which prizes are offered for the most similar pairs. Although identical genes do indeed contribute to a similarity between them, it is the pathological compulsion of their parents to create an inhuman identity between them that is most threatening to the individuality of genetically identical individuals. But even the most extreme efforts to turn genetic clones into human clones fail. The notion of "cloning Einstein" is a biological absurdity.

The error that supposes that organisms are determined totally by their genes is not simply a popular misunderstanding of biology. It is built into the structure of experiment and the theoretical basis of modern developmental biology as it is practiced. Developmental biology is an elaborate system of metaphorical filters through which all the observations of nature and experiment are screened. Development is the realization of an innate program, set by the genes. That program consists of a well-ordered set of stages that follow each other, each serving as a trigger for the next stage. Variation in final outcome is a consequence of individuals' premature arrest before the completion of the last stage. So the study of development is ultimately a study of the program itself and the genes that specify it. If one had a large enough computer and the complete sequence of an organism's DNA, one ought to be able to compute the organism from the program. The question is, What is the relationship between the actual observations and this metaphor? In particular, what has been left out of the description of development in order to cram the facts into such a theoretical framework?

First, we need to consider whether it is true that if one had a large enough computer and the DNA sequence of an organism, one could compute the organism. What do biologists know about the relationship between the development of organisms and their genes? Every biologist knows that development is contingent on environment, but that contingency is described in terms of potential (another metaphor). Some organisms have higher potential and some have lower potential. If they are given a good environment, they will all grow to their best potential, and then the intrinsic differences will be manifest. If they are given a deprived environment, then their potential will not be realized and they will all be equally immature and incomplete, so you will not be able to tell the difference in the intrinsic properties. The metaphor is of the empty bucket. If one's innate bucket is large, then it can hold a large amount poured into it, but small buckets will simply overflow. If only a small amount is poured into the buckets, then of course all will hold the same amount. This is the view of the development of intelligence, for example. If you give people really good education and good conditions for psychic development, then there are going to be immense differences between individuals in their performance because they start out with different potentials. If all you really want is equality, in this view, the way to get it is to deprive all children of a decent environment, and they will all be stupid together.

In fact, the empty bucket is a very bad metaphor for development. The correct concept is the norm of reaction transformation or table of correspondence between the environments in which organisms develop and the eventual phenotype. Figure 2.1 shows a norm of reaction of eye size in *Drosophila melanogaster* as a function of temperature. On the

Figure 2.1. Observed relationship between eye size and developmental temperature in three genotypes of *Drosophila*. (Courtesy of the author.)

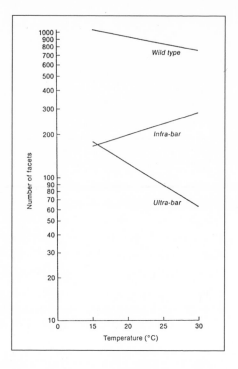

abscissa is developmental temperature from 15°C through 30°C. On the ordinate is the number of facets in the compound eye of a fly developing at each temperature. There are three norms of reaction shown, corresponding to three different genotypes. One is a genetically normal *Drosophila* showing that the number of eye facets gets smaller as the developmental temperature rises. The second norm of reaction is for a mutant called ultrabar that has fewer eye facets than normal at all temperatures. This norm is differently sensitive to temperature, but one can never confuse an ultrabar with a normal fly. Using the concept of potential, one would say that ultrabar flies have a lower potential for eye facet formation than normal flies. Irrespective of the environment, this is a clear difference. However, there is a third genotype on the graph, a mutant called infrabar that also has fewer eye facets. It is also always distinguishable from normal no matter what the temperature, but note that it is not always distinguishable from ultrabar. At high temperatures, infrabar has more facets than ultrabar, but at low temperatures it has fewer. The norms of reaction cross. The question is, Which of these patterns is typical of organisms and their development? Is it most of the time like the difference between normal and ultrabar, in which case we do not have to worry about the environment because the genes really tell us about capacity? Or is it usually like the contrast between ultrabar and infrabar, so that the curves go in different directions as a function of environment and cross each other? What is development like, not for mutants, but for the run of naturally occurring genotypes?

Figure 2.2 is taken from an experiment by Clausen, Keck, and Hiesey[1] on a plant, *Achillea*, from California. Mature, growing *Achillea* plants were cut in three pieces, and one piece of each plant was grown at low, one at medium, and one at high elevation. What is shown vertically in figure 2.2 is the same plant (in a genetic sense, because it was

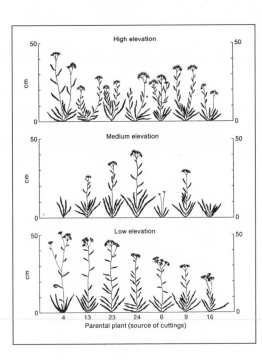

Figure 2.2. Growth of seven different plant genotypes in three environments. (Reprinted from J. Clausen, D. D. Keck, and W. W. Hiesey, *Experimental Studies on the Nature of Species,* vol. 1: *The Effect of Varied Environments on Western North American Plants,* No. 520 [Washington, D.C.: Carnegie Institute of Washington, 1940], 310 [fig. 122]. Reprinted with permission of the Carnegie Institute.)

cut into three pieces each with the same genes) regrowing in three different environments. The same "genetic program" has been given three chances to develop, like identical triplets. We note that there is a great deal of variation from plant to plant in how well they grow at a given elevation (horizontal comparisons in fig. 2.2). But we also notice that there is no relationship at all between how well a plant does at one elevation and how well it does at another. For example, the best growing plant, which flowered at the low elevation, barely grew at the medium elevation and did not flower. It was again the best at high elevation. The third plant grew nearly as well as the first at low elevation but was the second smallest at the higher elevation. The seventh plant did fairly poorly everywhere. There is no correlation between the performance of plants at one elevation and performance at others.

Yet another illustration of the development of "normal" genotypes is given in figure 2.3, showing the effect of temperature of development on the probability of survivorship of *Drosophila* larvae. The reaction norm for many different genotypes from a natural population is shown. A few fare poorly at all temperatures, but none is really superior at every temperature. One survives well at low temperature, reasonably well at intermediate temperature, and poorly at high temperature. Another goes up and down in viability, and another is high at low temperature, goes down, and comes back up at high temperature. In fact, there is no rule.

The metaphor of innate capacity is the wrong metaphor. There are differences among genotypes, with different consequences in different environments, but there is no way, over environments, to rate those innate or intrinsic properties from "bad" to "good," "high" to "low," "small" to "big." There is complete environmental contingency. No one knows whether human psychic development is like that, but if it is, then changes in the conditions of psychic development will produce very surprising results. All that can be said is that where norm-of-reaction studies have been done—in plants, fruit flies, and mammals—they all look the same. They all have the property that the norms cross each other, as in figures 2.2 and 2.3. Unlike in hybrid corn, mice, or flies, norms of reaction for psychic development will be very hard to mea-

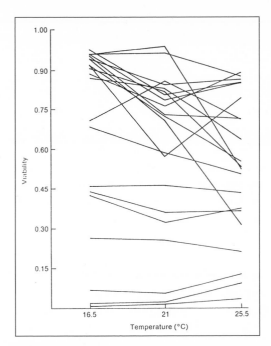

Figure 2.3. Relationship between probability of survival and developmental temperature for several genotypes taken from natural populations of *Drosophila*. (Courtesy of the author.)

sure. What is the environment? How do we measure the outcome of psychic development? And finally, most difficult of all, how do we observe the psychic development of the same genotype in different environments when all human beings (except identical twins and triplets) are genetically different? Can we really raise large numbers of identical twins in unrelated environments from birth?

Obviously, there are some genetic differences that appear in all normal environments. The children of blue-eyed, blonde English colonial administrators still were blue-eyed and blonde even though they were born and raised in early childhood in India. On the other hand, the descendants of Irish, Italian, Greek, and eastern European immigrants in Boston all speak an American English that bears the traces of their social class and education, but not of the genes of their ancestors. That is, some traits, such as the phonemic structure of human speech, appear to owe all their variation to environment and none to genetic difference. The contingency of organismic variation on genetics and environmental differences varies from trait to trait and from species to species, and there is no absolute law. All that we have is the empirical evidence from the species and traits that have been studied with a view to understanding the problem.

We can summarize the issues of developmental contingency considering three kinds of models of organism and environment shown in figure 2.4. Figure 2.4a is a model in which genes determine everything. Environments are not all the same, but plan A produces organism A and plan B produces organism B, irrespective of what comes from the environment. This is the view of the hegemony of the genes. It is the problematic of developmental biology that asks why lions give rise to lions and lambs give rise to lambs, irrespective of whether it is hot, cold, wet, or dry. It is the model of development and gene action that is in most textbooks. Figure 2.4b is exactly the inverse of figure 2.4a. Here everything depends on the environment and the genes only make the general rules of development. Environment A causes one kind of organism and environment B another. All differences depend on the environment, although genes are a necessary common cause of development. Both schemes are wrong as generalizations, although each one can be right for certain circumstances. There is no temperature at which one can turn a stone into a chicken. For the differences between lambs and lions, figure 2.4a is a good approximation. On the other hand, irrespective of what language one's immigrant ancestors spoke, we talk the same sort of flat trans-Atlantic English, irrespective of our genes. We need genes to speak, but given that one can speak, genes do not influence the differences in language between us. In this case, all normal genotypes are equivalent, as in figure 2.4b. But, in general, the world is much closer to figure 2.4c. Different genes and different environments, through unique interactions, give rise to unique organisms. To a first approximation, organisms are the consequence of genotype and environment in unique interaction, which is historical. That is, the order of environments matters. In fruit flies, if the temperature of development is changed, it matters whether it starts warm and then becomes cold or the reverse. What implication does that have for Head

Start Programs? The idea that it does not matter in what order human beings experience psychic environments cannot be right, in general. We will have to tailor carefully so-called remedial environments to be effective, taking into account the importance of environmental order in life history.

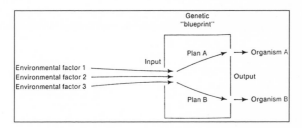

Figure 2.4A: A model of development in which genetic differences are determinative. (Courtesy of the author.)

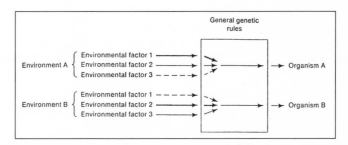

Figure 2.4B: A model of development in which environmental differences are determinative. (Courtesy of the author.)

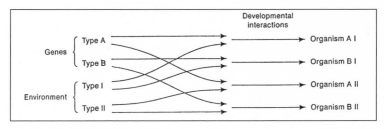

Figure 2.4C: A model of development in which both genes and environment account for differences. (Courtesy of the author.)

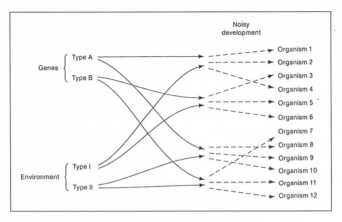

Figure 2.4D: A model of development that includes genetic and environmental differences and developmental noise. (Courtesy of the author.)

Yet even the complex picture in figure 2.4c is too naive. Pharmacologists experiment with inbred strains of mice bred to be genetically identical. These animals, housed in identical cages, are fed with exactly the same lab chow and water and then they are given some noxious chemical in various doses until they die. These genetically identical animals, living at the same temperature, all eating the same food, drinking the same water, do not all die at the same dosage of chemical. There is a dose-response curve such that as chemical concentration increases a few mice begin to die, then a few more, then 50 percent die at the LD_{50} dose. Why isn't the dose-response curve a step function? Why don't all mice drop dead at exactly the same concentration if they are genetically and environmentally identical?

A similar mystery appears in embryonic development. The fingerprints on a person's left hand are not identical to the fingerprints on the right hand, nor are the fingerprints of identical twins identical. The same genes are on the left and right sides, and hands develop in a rather homogeneous culture condition in utero. Why aren't left and right sides identical? Fruit flies have hairs on their bodies under their wings. The number is unequal on the left-hand and right-hand side of an individual fruit fly. There is on average over all flies an equal number, but some flies have more on the left side and some more on the right side. Nor can one establish a left-sided strain of flies by selective breeding. If one breeds only those with more hairs on the left, their offspring will still vary in the same way. The sides of a fly are genetically identical. A fruit fly is no bigger than the sharpened end of a lead pencil, and it develops, first, while burrowing in a semiliquid medium and, later, glued high up on the side of its culture vessel. We cannot conceive that the humidity, temperature, and CO_2 concentration are different on the left- and right-hand sides of that developing fly, yet a fly is very different on its left and right sides. There is as much difference in bristle number between left- and right-hand sides on individual flies as there is between flies that have different genotypes.

There is some unknown source of developmental variation within an organism that Conrad Waddington called developmental noise. Of course, we do not explain it by giving it a name. But noise is a fact of the life of all organisms in their development and is a very powerful source of variation between organisms as well as between sides of organisms. For many characteristics, the amount of variation generated by developmental noise is as great as from environmental and genetic differences. For bristles, we understand what the source of that noise is at the cellular level. Bristles are formed by single cells that divide into three. There is thermal noise at the molecular level that determines when cell division occurs and how long it takes. The cells destined to form bristles must migrate from deeper tissue layers into exterior layers. While this migration is occurring, the exterior cuticle is hardening, and if a bristle cell cluster arrives at the surface too late, the hardened cuticle will not allow a bristle to form. Thus, the thermal molecular noise becomes manifest as random variation in the presence or absence of a bristle.

Similar considerations may apply to psychic development. It may be that even if I had started to learn to play the violin at age four, I could not play like Perlman, who has nervous connections that I do not have. But to say that people have different nervous connections, even at birth, is not the same thing as saying that those differences in nervous connections are the consequence of genetic differences. We do not know how many of the connections in the central nervous system are a consequence of developmental noise. It is one of the leading problems of neural development. There are now random noise and selection theories of the development of the central nervous system[2] in which developmental noise can play a considerable role.

The organism does not compute itself from its genes, nor does it even compute itself from its genes and environment. Any computer that was as noisy as the developmental computer would be sent back to the factory. The correct model for development is given in figure 2.4d. It is the one that should be in all the textbooks. The metaphor for computation is the wrong metaphor for development. As a child I could not go to the movies or look at a picture magazine without being confronted by the genetically identical Dionne quintuplets, identically dressed and coiffed, on display in "Quintland" by Dr. Dafoe and the Province of Ontario for the amusement of tourists. This enforced homogenization continued through their adolescence, when they had been returned to their parents' custody. Yet each of their unhappy adulthoods was unhappy in its own way, and they seemed no more alike in career or health than we might expect from five girls of the same age brought up in a rural working-class French Canadian family. Three married and had families. Two trained as nurses, two went to college. Three were attracted to a religious vocation, but only one made it a career. One died in a convent at age twenty, suffering from epilepsy, one at age thirty-six, and three remain alive at sixty-three. So much for the doppelganger phenomenon.

Much is made of the moral responsibility of scientists to consider the social and humane implications of their increased power to manipulate the physical world. Far too little attention has been paid to the immense damage done by the propagation of false ideologies and false metaphors by scientists. What scientists say about the world is at least as important from a moral and political point of view as the actual state of nature.

NOTES

Parts of this essay have appeared in the *New York Review of Books* and in volume 20 of the Heinz Werner Lecture Series of Clark University. The author is extremely grateful to the Heinz Werner Lecture Committee for allowing him to reproduce that material here.

1. J. Clausen, D. D. Keck, and W. W. Hiesey, *Experimental Studies on the Nature of Species*, vol. 1: *The Effect of Varied Environments on Western North American Plants*, No. 520 (Washington, D.C.: Carnegie Institute of Washington, 1940), 1–452.

2. See J. Edelman, *Neural Darwinism* (New York: Basic Books, 1987).

PART 2

*Ethical Issues
in Cloning Humans*

3   There Will Never Be Another You

"Researchers Astounded" is not the typical phraseology of a headline on the front page of the *New York Times* (February 23, 1997). Lamb number 6LL3, better known as Dolly, took the world by surprise, sparking debate about the proper uses of biotechnology and inspiring predictable public fantasies (and predictable jokes). Recognizing that what is possible today with sheep will probably be feasible with human beings tomorrow, commentators speculated about the legitimacy of cloning Pavarotti or Einstein, about the chances that a demented dictator might produce an army of supersoldiers, and about the future of basketball in a world where the Boston Larry Birds play against the Chicago Michael Jordans. Polls showed that Mother Teresa was the most popular choice for person-to-be-cloned, although a film star (Michelle Pfeiffer) was not far behind, and Bill Clinton and Hillary Clinton obtained some support.

Mary Shelley may have a lot to answer for. The Frankenstein story, typically in one of its film versions, colors popular reception of news about cloning, fomenting a potent brew of associations: We assume that human lives can be created to order, that it can be done instantly, that we can achieve exact replicas, and, of course, that it is all going to turn out disastrously. Reality is much more sobering, and it is a good idea to preface debates about the morality of human cloning with a clear understanding of the scientific facts.

In 1975, developmental biologist John Gurdon reported the possibility of cloning amphibians. Gurdon and his coworkers were able to remove the nucleus from a frog egg and replace it with the nucleus from an embryonic tadpole. The animals survived and developed to adulthood, becoming frogs that had the same complement of genes within the nucleus as the tadpole embryo. Further efforts to use nuclei from adult donors were unsuccessful. When the nucleus from a frog egg was replaced with that from a cell taken from an adult frog, the embryo died at a relatively early stage of development. Moreover, nobody was able to perform Gurdon's original trick on mammals. Would-be cloners who tried to insert nuclei from mouse embryos into mouse eggs consistently ended up with dead fetal mice. So, despite initial hopes and fears, it appeared that the route to cloning adult human beings was doubly blocked: Only transfer of embryonic DNA seemed to work, and even that failed in mammals.

Biologists had an explanation for the failure to produce normal development after inserting nuclei from adult cells. Although adult cells contain all the genes, they are also *differentiated,* set to perform particular functions, and this comes about because some genes are expressed in them, while others are "turned off." Regulation of genes is a matter of the attachment of proteins to the DNA so that some regions are accessible for transcription and others are not. So it was assumed that chromosomes in adult cells would have a complex coating of proteins on the DNA and this would prevent the transcription of genes that need to be activated in early development. In consequence, transferring nuclei from adult cells always produced inviable embryos. The "high-tech" solution to the problem—the "obvious" solution from the viewpoint of molecular genetics—is to use the arsenal of molecular techniques to strip away the protein coating, restore the DNA to its (presumed) original condition, and only then transfer the nucleus to the recipient egg. To date, nobody has managed to make this approach work.

The breakthrough came not from one of the major centers in which the genetic revolution is whirling on, but from the far less glamorous world of animal husbandry and agricultural research. In 1996, a team of workers at the Roslin Institute near Edinburgh, led by Dr. Ian Wilmut, announced that they had succeeded in producing two live sheep, Megan and Morag, by transplanting nuclei from embryonic sheep cells. One barrier had been breached: Wilmut and his colleagues had shown that just what Gurdon had done in frogs could be achieved in sheep. Yet it seemed that the major problem, that of tricking an egg into normal development when equipped with an *adult* cell nucleus, still remained. In retrospect, we can recognize that not quite enough attention was given to Wilmut's first announcement, for Megan and Morag testified to a new technique of nuclear transference.

Wilmut conjectured that the failures of normal development resulted from the fact that the cell that supplied the nucleus and the egg that received it were at different stages of the cell cycle. Using well-known techniques from cell biology, he "starved" both cells, so that both were in an inactive phase at the time of transfer. In a series of experiments, he discovered that inserting nuclei from adult cells (from the udder of a pregnant ewe) under this regimen gave rise to a number of embryos, which could be implanted in ewes. Although there was a high rate of miscarriage, one of the pregnancies went to term. So, after beginning with 277 successfully transferred nuclei, Wilmut obtained one healthy lamb, the celebrated Dolly.

Wilmut's achievement raises three important questions: Will it be possible to perform the same operations on human cells? Will donors be able to reduce the high rate of failure? What exactly is the relationship between a clone obtained in this way and previously existing animals? Answers to the first two of these are necessarily tentative, since predicting even the immediate trajectory of biological research is always vulnerable to unforeseen contingencies. (In the weeks after Gurdon's success, it seemed that cloning all kinds of animals was just around the corner; from the middle 1980s to 1996, it appeared that cloning adult mammals was a science-fiction fantasy.) However, unless there is some quite unanticipated snag, we can expect that Wilmut's technique will *eventually* work just as well on human cells as it does in sheep, and that failure rates in sheep (or in other mammals) will quickly be reduced.

Assuming that Wilmut's diagnosis of the problems of mammalian cloning is roughly correct, then the crucial step involves preparing the

cells for nuclear transfer by making them quiescent. Of course, learning how to "starve" human cells so that they are ready may involve some
experimental tinkering. Probably there would be a fair bit of trial-and-
error work before the techniques became sufficiently precise to allow
embryos to develop to the stage at which they can be implanted with a
very high rate of success, and to overcome any potential difficulties with
implantation or with the resultant pregnancy. Many of the problems
that prospective human cloners would face are likely to be analogues
of obstacles to the various forms of assisted reproduction, and it is
perfectly possible that the successes of past human reproductive technology would smooth the way for cloning.

On the third question we can be more confident. Dolly has the same
nuclear genetic material as the adult pregnant ewe, from whose udder
cell the inserted nucleus originally came. A different female supplied
the egg into which the nucleus was inserted, and Dolly thus has the
same mitochondrial DNA as this ewe; indeed, her early development
was shaped by the interaction between the DNA in the nucleus and the
contents of the cytoplasm, the contributions of different adult females.
Yet a third sheep, the ewe into which the embryonic Dolly was implanted, played a role in Dolly's nascent life, providing her with a uterine
environment. In an obvious sense, Dolly has three mothers—nucleus
mother, egg mother, and womb mother—and no father (unless, of
course, we give Dr. Wilmut that honor for his guiding role).

Now imagine Polly, a human counterpart of Dolly. Will Polly be a
replica of any existing human being? Certainly she will not be the same
person as any of the mothers—even the nuclear mother. Personal identity, as philosophers since John Locke have recognized, depends on
continuity of memory and other psychological attitudes. There is no
hope of ensuring personal survival by arranging for cloning through
supplying a cell nucleus. Megalomaniacs with intimations of immor-
tality need not apply.

Yet you might think that Polly might be very similar to her nuclear
mother, perhaps extremely similar if we arranged for the nuclear mother to be the same person as the egg mother, and for that person's mother
to be the womb mother. That combination of "parents" would seem

to turn Polly into a close approximation of her nuclear mother's identical twin. An approximation, perhaps, but nobody knows how close. Polly and her nuclear mother differ in three ways in which identical twins are typically the same. They develop from eggs with different cytoplasmic constitutions, they are not carried to term in a common uterine environment, and their environments after birth are likely to be quite different.

Interestingly, during the next few years, Wilmut's technique will allow us to remedy our ignorance about the relative importance of various causes of phenotypic traits by performing experiments on nonhuman mammals. It will be possible to develop organisms with the same nuclear genes within recipient eggs with varied cytoplasms. By exploring the results, biologists will be able to discover the extent to which constituents of the egg outside the nucleus play a role in shaping the phenotype. They will also be able to explore the ways in which the uterine environment makes a difference. Perhaps they will find that variation in cytoplasm and difference in womb have little effect, in which case Polly will be a better approximation to an identical twin of her nuclear mother. More probably, I believe, they will expose some aspects of the phenotype that are influenced by the character of the cytoplasm or by the state of the womb, thus identifying ways in which Polly would fall short of perfect twinhood.

Even before these experiments are done, we know of some important differences between Polly and her nuclear mother. Unlike most identical twins, they will grow up in environments that are quite dissimilar, if only because the gap in their ages will correspond to changes in dietary fashions, educational trends, adolescent culture, and so forth. When these sources of variation are combined with the more uncertain judgments about effects of cytoplasmic factors and the prenatal environment, we can conclude that human clones will be less alike than identical twins, and quite possibly very much less alike. Those beguiled by genetalk move quickly from the idea that clones are genetically identical (which is, to a first approximation, correct) to the view that clones will be replicas of one another. Identical twins reared together are obviously similar in many respects, but they are by no means interchangeable people. It is pertinent to recall the statistics about sexual orientation: 50 percent of male (identical) twins who are gay have a co-twin who is not. Minute differences in shared environments can obviously play a large role. How much more dissimilarity can we anticipate given the much more dramatic variations that I have indicated?

There will never be another you. If you hoped to fashion a son or daughter exactly in your own image, you would be doomed to disappointment. Nonetheless, you might hope to use cloning technology to have a child of a particular kind—just as the obvious agricultural applications focus on single features of domestic animals, like their capacity for producing milk. Some human characteristics are under tight genetic control, and if we wanted to ensure that our children carried genetic diseases like Huntington's and Tay-Sachs, then, of course, we could do so, although the idea is so monstrous that it only surfaces in order to be dismissed. Perhaps there are other features that are relatively insusceptible to niceties of the environment, aspects of body morphology, for example. An obvious example is eye color.

Imagine a couple determined to do what they can to produce a Hollywood star. Fascinated by the color of Elizabeth Taylor's eyes, they obtain a sample of tissue from the actress and clone a young Liz. For reasons already discussed, it is probable that Elizabeth II would be different from Elizabeth I, but we might think that she would have that distinctive eye color. Supposing that to be so, should we conclude that the couple will realize their dream? Probably not. Waiving issues about intelligence, poise, and acting ability and supposing that the movie moguls of the future respond only to physical attractiveness, the eyes may not have it. Apparently tiny incidents in early development may modify the shape of the orbits, producing a combination of features in which the eye color no longer has its bewitching effect. At best, the confused couple can only hope to raise the probability that their daughter will capture the hearts of millions.

Physical attractiveness, the real target of the couple's plan, turns on more than eye color, and that is the general way of things. The traits we value most are produced by a complex interaction between genotypes and environments, and by fixing the genotype, we can only increase our chances of achieving the results we want. Demented dictators bent on invading their neighbors can do no more than add to the likelihood of generating the "master race." Before we startle ourselves with the imagined sound of jackboots marching across the frontier, we should remember that there is no shortcut to the process of rearing children and training them in whatever ways are appropriate to our ends. Indeed, when we appreciate that point, we can see that if the dictators are slightly less demented, they will do what military recruiters have always done, namely select on grounds of physical fitness, ease of indoctrination, courage, and such traits and then invest extensively in military academies. Cloning adds very little to the chances of success.

Similar points apply to the fantasies of cloning Einstein, Mother Teresa, or Yo Yo Ma. The chances of generating true distinction in any area of complex human activity, whether it be scientific accomplishment, dedication to the well-being of others, or artistic expression, are infinitesimal. *Perhaps* cloning would allow us to raise the probability from infinitesimal to very, very tiny. A program designed to use cloning to transform human life by having a higher number of outstanding individuals would, at most, give a minute number of "successes" at the cost of vastly more "failures." Those who worry that Dolly is one survivor among 277 attempts should find this scenario far more disturbing.

Garish popular fantasies dissolve when confronted with the facts about the biotechnology of cloning, suggesting that only rich recluses, hermetically sealed in ignorance, should be tempted by the projects that fascinate and horrify us most. Yet there are other more mundane ventures that have a closer connection with reality. Parents who demand less than truly outstanding performance, but still have a preferred dimension on which they want their children to excel, might turn to cloning in hopes of raising their chances. Had my wife and I been seriously concerned to bring into the world sons who would have dominated the basketball court or been mainstays of the defensive line, then we would have been ill-advised to proceed in the old-fashioned method of reproduction. At a combined weight of less than 275 pounds and a combined height of just over 11 feet, we would have done far better to transfer a nucleus from some strapping star of the NBA or the NFL.

Perhaps, by doing so, we would significantly have raised the chances of having a son on the high-school basketball or football team. Success, even at that rather modest goal, would not have been guaranteed, since there are all kinds of ways in which the boy's development might have gone differently (think of accidents, competing interests, dislike of competitive sports, and so forth). Nor would cloning necessarily have been the best way for us to proceed: Maybe we could have employed the results of genetic testing to produce, by in vitro fertilization, a fertilized egg having alleles at crucial loci that predispose to a large, muscular body; maybe we could have used artificial insemination, or have adopted a son. Nevertheless, cloning would surely have raised the probabilities of our obtaining the child we wanted.

Just that final phrase indicates the moral squalor of the story. As I have imagined it, we have a plan for the life to come laid down in advance; we are determined to do what we can to make it come out a certain way, and, presumably, if it does not come out that way, it will count as a failure. Throughout the discussion of utopian eugenics, I insisted that prenatal decisions should be guided by reflection on the quality of the nascent life, and I understood that in terms of creating the conditions under which a child could form a central set of desires, a conception of what his or her life means that had a decent chance of being satisfied. In the present scenario, there is a crass failure to recognize the child as an independent being, one who should form his own sense of who he is and what his life means. The contours of the life are imposed from without.

Parents have been tempted to do similar things before. James Mill had a plan for his son's life, leading him to begin young John Stuart's instruction in Greek at age three and Latin at age eight. John Stuart Mill's *Autobiography* is a quietly moving testament to the cramping effect of his felt need to live out the life his eminent father had designed for him. In early adulthood, Mill *fils* suffered a nervous breakdown, from which he recovered, going on to a career of great intellectual distinction. Although John Stuart partially fulfilled his father's aspirations for him, one of the most striking features of his philosophical work is his passionate defense of human freedom. The central point about what was wrong with this father's attitude toward a son has never been better expressed than in the splendid prose of *On Liberty:* "Mankind are greater gainers by suffering each other to live as seems good to themselves than by compelling each to live as seems good to the rest."

If cloning human beings is undertaken in the hope of generating a particular kind of person, a person whose standards of what matters in life are imposed from without, then it is morally repugnant, not because it involves biological tinkering but because it is continuous with other ways of interfering with human autonomy that we ought to resist. Human cloning would provide new ways of committing old moral errors. To discover whether or not there are morally permissible cases of cloning, we need to see if this objectionable feature can be removed, if there are situations in which the intention of the prospective parents is properly focused on the quality of human lives but in which cloning represents the only option for them. Three scenarios come immediately to mind.

1. *The case of the dying child.* Imagine a couple whose only son is slowly dying. If the child were provided with a kidney transplant within the next ten years, he would recover and be able to lead a normal life. Unfortunately, neither parent is able to supply a compatible organ, and it is known that individuals with kidneys that could be successfully transplanted are extremely rare. However, if a brother were produced by cloning, then it would be possible to use one of his kidneys to save the life of the elder son. Supposing that the technology of cloning human beings has become sufficiently reliable to give the couple a very high probability of successfully producing a son with the same complement of nuclear genes, is it permissible for them to do so?

2. *The case of the grieving widow.* A woman's much-loved husband has been killed in a car crash. As the result of the same crash, the couple's only daughter lies in a coma, with irreversible brain damage, and she will surely die in a matter of months. The widow is no longer able to bear children. Should she be allowed to have the nuclear DNA from one of her daughter's cells inserted in an egg supplied by another woman, and to have a clone of her child produced through surrogate motherhood?

3. The case of the loving lesbians. A lesbian couple, devoted to one another for many years, wish to produce a child. Because they would like the child to be biologically connected to each of them, they request that a cell nucleus from one of them be inserted in an egg from the other, and that the embryo be implanted in the woman who donated the egg. (Here, one of the women would be nuclear mother and the other would be both egg mother and womb mother.) Should their request be accepted?

In all of these instances, unlike the ones considered earlier, there is no blatant attempt to impose the plan of a new life, to interfere with a child's own conception of what is valuable. Yet there are lingering concerns that need to be addressed. The first scenario, and to a lesser extent the second, arouses suspicion that children are being subordinated to special adult purposes and projects. Turning from John Stuart Mill to one of the other great influences on contemporary moral theory, Immanuel Kant, we can formulate the worry as a different question about respecting the autonomy of the child: Can these cases be reconciled with the injunction "to treat humanity whether in your own person or in that of another, always as an end and never as a means only"?

It is quite possible that the parents in the case of the dying child would have intentions that flout that principle. They have no desire for another child. They are desperate to save the son they have, and if they could only find an appropriate organ to transplant, they would be delighted to do that; for them the younger brother would simply be a cache of resources, something to be used in saving the really important life. Presumably, if the brother were born and the transplant did not succeed, they would regard that as a failure. Yet the parental attitudes do not have to be so stark and callous (and, in the instances in which parents have actually contemplated bearing a child to save an older sibling, it is quite clear that they are much more complex). Suppose we imagine that the parents plan to have another child in any case, that they are committed to loving and cherishing the child for his or her own sake. What can be the harm in planning that child's birth so as to allow their firstborn to live?

The moral quality of what is done plainly depends on the parental attitudes, specifically on whether or not they have the proper concern for the younger boy's well-being, independently of his being able to save his elder brother. Ironically, their love for him may be manifested most clearly if the project goes awry and the first child dies. Although that love might equally be present in cases where the elder son survives, reflective parents will probably always wonder whether it is untinged by the desire to find some means of saving the first-born—and, of course, the younger boy is likely to entertain worries of a similar nature. He would by no means be the first child to feel himself a second-class substitute, in this case either a helpmeet or a possible replacement for someone loved in his own right.

Similarly, the grieving widow might be motivated solely by desire to forge some link with the happy past, so that the child produced by cloning would be valuable because she was genetically close to the dead (having the same nuclear DNA as her sister, DNA that derives from the widow and her dead husband). If so, another person is being treated as a means to understandable, but morbid, ends. On the other hand, perhaps the widow is primarily moved by the desire for another child, and the prospect of cloning simply reflects the common attitude of many (though not all) parents who prize biological connection to their offspring. However, as in the case of the dying child, the participants, if they are at all reflective, are bound to wonder about the mixture of attitudes surrounding the production of a life so intimately connected to the past.

The case of the loving lesbians is the purest of the three, for here we seem to have a precise analogue of the situation in which heterosexual couples find themselves. Cloning would enable the devoted pair to have a child biologically related to both of them. There is no question of imposing some particular plan on the nascent life, even the minimal one of hoping to save another child or to serve as a reminder of the dead, but simply the wish to have a child who is their own, the expression of their mutual love. If human cloning is ever defensible, it will be in contexts like this.

During past decades, medicine has allowed many couples to overcome reproductive problems and to have biological children. The development of techniques of assisted reproduction responds to the sense that couples who have problems with infertility have been deprived of something that it is quite reasonable for people to value, and that various kinds of manipulations with human cells are legitimate responses to their frustrations. Yet serious issues remain. How close an approximation to the normal circumstances of reproduction and the normal genetic connections should we strive to achieve? How should the benefits of restoring reproduction be weighed against possible risks of the techniques? Both kinds of questions arise with respect to our scenarios.

Lesbian couples already have an option to produce a child who will be biologically related to both. If an egg from one of them is fertilized with sperm (supplied, say, by a male relative of the other) and the resultant embryo is implanted in the womb of the woman who did not give the egg, then both have a biological connection to the child (one is egg mother, the other womb mother). That method of reproduction might even seem preferable, diminishing any sense of burden that the child might feel because of special biological closeness to one of the mothers and allowing for the possibility of having children of either sex. The grieving widow might turn to existing techniques of assisted reproduction and rear a child conceived from artificial insemination of one of her daughter's eggs. In either case, cloning would create a closer biological connection—but should that extra degree of relationship be assigned particularly high value?

My discussion of all three scenarios also depends on assuming that human cloning works smoothly, that there are no worrisome risks that the pregnancy will go awry, producing a child whose development is seriously disrupted. Dolly, remember, was one success out of 277 tries, and we can suppose that early ventures in human cloning would have an appreciable rate of failure. We cannot know yet whether the development of technology for cloning human beings would simply involve the death of early embryos, or whether, along the way, researchers would generate malformed fetuses and, from time to time, children with problems undetectable before birth. During the next few years, we shall certainly come to know much more about the biological processes involved in cloning mammals, and the information we acquire may make it possible to undertake human cloning with confidence that any breakdowns will occur early in development (before there is a person with rights). Meanwhile, we can hope that the continuing transformation of our genetic knowledge will provide improved methods of transplantation, and thus bring relief to parents whose children die for lack of compatible organs.

Should human cloning be banned? Until we have much more extensive and detailed knowledge of how cloning can be achieved (and what the potential problems are) in a variety of mammalian species, there is no warrant for trying to perform Wilmut's clever trick on ourselves. I have suggested that there are some few circumstances in which human cloning might be morally permissible, but, in at least two of these, there are genuine concerns about attitudes to the nascent life, while, in the third, alternative techniques, already available, offer almost as good a response to the underlying predicament. Perhaps, when cloning techniques have become routine in nonhuman mammalian biology, we may acknowledge human cloning as appropriate relief for the parents of dying children, for grieving widows, and for loving lesbians. For now, however, we do best to try to help them in other ways.

Dolly, we are told, like the scientist who helped her into existence, is learning to live with the television cameras. Media fascination with cloning plainly reached the White House, provoking President Clinton first to refer the issue to his newly formed Bioethics Advisory Committee, later to ban federal funding of applications of cloning technology to human beings. The February 27, 1997, issue of *Nature* featuring Dolly offered a less-than-positive assessment of the presidential reaction: "At a time when the science policy world is replete with technology foresight exercises, for a U.S. president and other politicians only now to be requesting guidance about what appears in today's *Nature* is shaming."

At the first stages of the Human Genome Project, James Watson argued for the assignment of funds to study the "ethical, legal, and social implications" of the purely scientific research. Watson explicitly drew the analogy with the original development of nuclear technology, recommending that, this time, scientific and social change might go hand in hand. Almost a decade later, the mapping and sequencing are advancing faster than most people had anticipated—and the affluent nations remain almost where they were in terms of supplying the social backdrop that will put the genetic knowledge to proper use. That is not for lack of numerous expert studies that outline the potential problems and propose ways of overcoming them. Much has been written. Little has been done. In the United States we still lack the most basic means of averting genetic discrimination, to wit universal health coverage, but Britain and even the continental European nations are little better placed to cope with what is coming.

The moral problems raised by the possibility of human cloning should be addressed by drawing on general moral principles, articulated in many contexts and in many idioms in the history of thought. Those principles advance a conception of what matters in human life, of what proper attitudes toward others should be, and, more specifically, of how we should treat nascent lives. Once the factual confusions about cloning are cleared away and once we have appealed to broader moral theory, we can see how to navigate our way through the territory and which possibilities are especially difficult. However, the moral principles cannot be applied selectively, nor can we dodge their implications either within the domain of uses of biotechnology or in thinking about our duties to children who are deprived of genuine opportunities for health and happiness. Only moral chameleons call for a ban on human cloning because of remote potential harms while instituting or supporting policies that permit children to live without proper health care and even endanger their prospects of food and shelter.

The belated response to cloning is of a piece with a general failure to translate clear moral directives into regulations and policies. Dolly is a highly visible symbol, but behind her is a broad array of moral issues that citizens of affluent societies seem to prefer to leave in the shadows. However strongly we feel about the plight of loving lesbians, grieving widows, or even couples whose children are dying, deciding the legitimate employment of human cloning in dealing with their troubles is not our most urgent problem. Those who think that working out the proper limits of human cloning is the big issue are suffering from moral myopia.

General moral principles provide us with an obligation to improve the quality of human lives, where we have the opportunity to do so, and developments in biotechnology provide opportunities and challenges. If we took the principles seriously, we would be led to demand serious investment in programs to improve the lives of the young, the disabled, and the socially disadvantaged. That is not quite what is going on in the "civilized" world. Making demands for social investment seems quixotic, especially at a time when, in America, funds for poor children and disabled people who are out of work are being slashed, and when, in other affluent countries, there is serious questioning of the responsibilities of societies to their citizens. Yet the application of patronizing adjectives does nothing to undermine the legitimacy of the demands. What is truly shameful is not that the response to possibilities of cloning came so late, nor that it has been confused; it is the common reluctance of all the affluent nations to think through the implications of time-honored moral principles and to design a coherent use of the new genetic information and technology for human well-being.

NOTE

This chapter previously appeared as pp. 327–42 of *The Lives to Come,* © 1997 by Philip Kitcher, and is reprinted with the permission of Simon & Schuster.

4   Cloning Human Beings:
      Sorting through the Ethical Issues

In 1997, just ten days after Dr. Ian Wilmut announced that he had successfully cloned a lamb using a technique known as somatic cell nuclear
transfer (SCNT), President Clinton imposed a temporary moratorium
on federal funding of human cloning research. He then asked the National Bioethics Advisory Commission (NBAC) to undertake a thorough
review of the legal and ethical issues associated with cloning and to report back to him within ninety days with recommendations on possible federal actions to prevent its abuse. The president said, "Each human
life is unique, born of a miracle that reaches beyond laboratory science.
. . . I believe we must respect this profound gift and resist the temptation
to replicate ourselves."[1] In light of his words, NBAC's recommendation
that the president's moratorium be continued was a surprise to no one.
Realistically, a moratorium on federal funding of human cloning research
was unnecessary because such research is already covered by a 1994 presidential ban on virtually all embryo research. In addition, various mechanisms are already in place to prevent unethical or inappropriate research,
such as institutional review boards (IRBs). It is highly unlikely that a
proposal to conduct experiments in human cloning would get past an
IRB, if only because the necessary animal research has not yet been done.
However, all of this is *politically* irrelevant. NBAC had to endorse the president's moratorium not to prevent the real possibility of abuse, but to
express in symbolic fashion the opposition to a technology viewed by
many people as profoundly disturbing.

At the same time, NBAC did not offer a permanent ban. Nor, to the disappointment of many commentators, were its recommendations based on specifically moral or religious objections to cloning, such as the fear that it would undermine human individuality and dignity, encourage the treating of children as commodities, or threaten the integrity of families. Instead, its recommendation was based virtually entirely on safety considerations, such as the high probability of failure and consequent high risk of miscarriage and an unknown risk of developmental abnormalities in offspring. Safety considerations are certainly relevant to the permissibility of research, but one scarcely needs a national bioethics commission to point that out. Why, then, did the commission base its endorsement of a moratorium on safety considerations alone? Did the commission, as some critics allege, duck the moral issues? Not at all. The commission listened to, and presented in its report, a wide range of moral and religious objections, but it took no stance on them, for the simple reason that it could not reach consensus in the limited time it had.

Even if the commission had considerably longer than ninety days, it probably could not have reached consensus on the moral claims presented to it. There are profound differences of opinion on the ethics of cloning human beings, which reflect fundamental differences in values and attitudes. However, rather than metaphorically throw up our hands in the face of these differences, I maintain that we can make progress by sifting through the various ethical arguments that have been made, evaluating them for factual accuracy, logical consistency, and persuasiveness. Some of the objections to human cloning can be dispensed with fairly easily because they are based on factual errors, misunderstanding, or fallacies. Others are more sophisticated and express deeper concerns. Sometimes it is difficult to articulate the moral misgivings people have when confronting a new technology. Nevertheless, it is important to try to do so if we are to get beyond purely visceral reactions that are not universally shared.

Playing God

In his charge to NBAC, President Clinton warned against the temptation to "play God." The correct response to this objection was made by Immanuel Jakobovits, former chief rabbi of Great Britain, who said, "We can dismiss the common argument of 'playing God' or 'interfering with divine providence.' Every medical intervention represents such interference."[2] When we give people new kidneys, or restart their hearts, or even vaccinate children, we are saving people who otherwise would have died. When we enable infertile couples to have children by in vitro fertilization, egg donation, or intracytoplasmic sperm injection (ICSI), we are playing God. Unless we reject all medical interventions, there is no principled way to decide which ones are objectionable because they are "playing God."

However, there is a reasonable interpretation of the "playing God" argument: It can be seen as a warning against hubris. In our enthusiasm for the potential benefits of a new technology, we may overlook the potential dangers. We may underestimate the risks or think that we have foreseen or can prevent all the harms. But any medical intervention can have unintended and disastrous results. Robert Pollack, professor of biology at Columbia University, makes the point this way:

> Every new technology is imperfect. As anyone knows who has tripped over the newest model of a computer or a car, the first tries are likely to have hidden flaws. This has been true of medical technology as well: the first vaccines, the first antibiotics, and the first organ transplants all had dangerous, if temporary, side effects. . . .
>
> Recently, for example, scientists interested in coloring the hair and eyes of an albino strain of mice injected the gene for a pigment; unexpectedly, they created a strain of mice whose viscera—heart, stomach, liver, and the like—were all turned around. These mice were unable to live long after birth; the added gene had inadvertently damaged a gene responsible for the usual positioning of the internal organs.[3]

It is worth reminding scientists and policymakers of the risks of unforeseen, harmful side effects. However, this interpretation of the "playing God" objection is simply a restating of safety concerns, which bear on any novel technology, and not a special objection to cloning.

The Threat to Individuality

Some people object to the creation of a new human being by cloning an already existing human being because they think that the clone would be a replica of the original person, alike in all respects, and hence not a unique individual. Possibly this idea stems from the fact that the first clone was a sheep, a species not noted for individuality. As the humorist Dave Barry expressed it, do we really need scientists spending time and money trying to make sheep *more* identical? Or perhaps the idea that human cloning threatens individuality stems from a perception, encouraged by joking comments in the media, that cloning would create a literal double, that, for example, if you cloned Bill Clinton, you'd get a middle-aged man, with a passion for politics and Big Macs, who could be negotiating peace in the Middle East and meeting with the Cabinet in the White House at the same time. Of course, this is nonsense. Cloning doesn't create a duplicate of the adult, but rather an infant, and that infant would no more grow up to be a replica of Bill Clinton than would his identical twin. As anyone who knows a pair of identical twins can attest, they're not exactly alike. They have different interests and different abilities, and they don't even look exactly alike. Indeed, animals created by SCNT cloning would be even less alike than identical twins in nature because they would not be genetically identical (due to the small genetic contribution from the mitochondrial DNA of the egg) and because the clone would be gestated either in a different uterus or in the same uterus at a different time. These gestational influences could also be expected to influence what the resulting person was like. Therefore, cloning does not destroy individuality or uniqueness.

The Child's Right to an Open Future

A related fear is that the future of a cloned individual would be predetermined, narrowing the child's choices when he or she grows up. This violates the child's "right to an open future," as Joel Feinberg expresses it,[4] or the child's "right to ignorance," as Hans Jonas puts it.[5] Whereas Feinberg did not expressly consider cloning as potentially violating the right to an open future, Jonas reflected on the dangers of cloning to self-development, years before SCNT was even a possibility. Jonas argued that each of us develops a personality and becomes a self by making choices. However, a cloned human being would know the choices that were made by the person whose genome he or she shared. In this way, a clone would differ from an identical twin, as identical twins go through life at the same time. By contrast, a clone would be a genetic replica of someone who has already lived his life, so the clone would know a great deal about himself and his future. He would know what he would look like as an adult, the diseases to which he would be prone, the talents he would have, and so forth. Thus, he would be unable to create and become his own self.

However, this objection is based on a fallacious assumption: that if you know what your genome is, you will know what your choices, and hence your life, will be. This is the fallacy of genetic determinism, aptly exposed in the NBAC Report. To put it bluntly, we are not our genes, and our genes do not determine what we are or will be. In the words of the report, "Each individual is . . . the result of a complex interaction between his or her genes and the environment within which they develop. . . . Indeed, the great lesson of modern molecular genetics is the profound complexity of both gene-gene interactions and gene-environment interaction in the determination of whether a specific trait or characteristic is expressed. In other words, there will never be another you."[6] A clone might have a pretty good idea what she would look like at age fifty; she would have considerable knowledge about the diseases she was at risk for, perhaps more than nonclones. However, she could not assume, any more than offspring conceived the usual way, that she would have the same abilities and talents as her progenitor, for this depends at least as much on environmental factors as on genetic inheritance.

Free Will

Even intelligent and educated people have expressed the fear that cloning would enable evil dictators to create armies of mindless beings, zombies, whom they could bend to their will. I used to think that this fear stemmed from an indiscriminate consumption of science fiction books and movies, from *The Boys from Brazil* to *Blade Runner*. However, this fear may also be the result of our not-so-distant history. In her book *Clone,* Gina Kolata quotes Alta Charo, a member of NBAC, as saying,

> The most frightening footages from World War II are not the ones that show the skeletal bodies of the survivors and not even the mushroom clouds over Hiroshima and Nagasaki. The most frightening are the crowds of Germans, with their fists raised, shouting Sieg Heil. The most frightening is the mob psychology that reduces individuals to nothing but clones of each other . . . people who can be easily manipulated, who can become an unthinking mass that can be a force of oppression.[7]

That fear is at the heart of the fear of cloning. It is not a silly fear. Nevertheless, is there any reason to think that a human being who was created by cloning would be less free, less autonomous, more susceptible to manipulation than people created the ordinary way? If a human being could be created by SCNT technology (something we have no way of knowing at this point), there is no reason to think that he or she would be any less free or autonomous than the rest of us. Why should the technique for getting an egg to act like an embryo affect the free will of the resulting human being?

Perhaps the fear is that cloning would allow for genetic manipulation, enabling the deliberate creation of people of low intelligence who could be manipulated into doing the bidding of others. In the hands of an unscrupulous dictator, these cloned individuals could become a tool of oppression. However, this is not a very realistic possibility in light of the difficulty, expense, and length of time required to produce followers by human cloning. If we are seriously worried about the creation of mindless followers of dictators, it would make more sense to focus on the economic and social conditions that conduce to mob psychology.

Exploitation of Cloned Individuals

The combination of cloning with genetic manipulation has been condemned because it could lead to exploitation of cloned individuals. For example, individuals of low intelligence might be deliberately

created to do repetitive, boring, and low-paying jobs. Their low intelligence would make them unsuitable for any other work; perhaps they
would not even mind jobs that most of the rest of us would avoid. If
they are of sufficiently low intelligence, it might not be possible for
them to mind. Whether this is genuinely exploitive, given that they do
not mind the work, is a topic far beyond the scope of this chapter.[8] But
if my sense that this would count as exploitation is correct, what's
wrong is not the technology itself, but the use to which it was put. The
fact that cloning *could* be used to create people of low intelligence in
order to exploit them is not a reason to object to cloning because any
technology can be used in wrongful ways. It would have to be shown
that the technology would inevitably be used that way and that the
exploitation could not be prevented.

In my view, the objections to human cloning just raised are met
fairly easily. Let us now turn to objections that raise questions that are
more difficult to resolve.

The Fear of Eugenics

If genetic manipulation can produce individuals of low intelligence,
presumably it could also be used to create "superior people," people
who are smarter, stronger, healthier, or more attractive. (This is a big
"if" because these are all complex, multifactorial traits. It is unlikely that
we will ever be able to determine through genetic intervention a person's level of intelligence, but it is possible that genetic intervention
might have an important influence.) Because such people would not
have reduced opportunities, but rather expanded ones, such manipulation would not be exploitive. Nevertheless, some people object to any
attempt to improve human beings via genetic manipulation. This is
eugenics, they say, and eugenics is just wrong.

However, this objection must be examined more closely. What exactly is objectionable about eugenics? Undoubtedly, some of the opposition stems from the history of the eugenics movement. Many early twentieth-century eugenicists held racist and classist beliefs; being
of non–Anglo Saxon ancestry was considered a form of hereditary

defect.[9] Blacks, Native Americans, Jews, and dark-skinned European ethnic groups were considered inferior "races." Prejudice against those with mental or physical impairments was rampant. "Mental defectives" were viewed as potential criminals. Birth control, sterilization, abortion, and sometimes even infanticide were advocated to protect society from what were called "lives of no value."[10] In Germany, eugenics played an important role in Nazi ideology and the legitimation of Nazi genocide.

However, rejection of a eugenics based on racist prejudice does not commit us to the view that *any* attempt to manipulate the genome or improve our genetic inheritance is necessarily bad. Manipulating the genome to prevent lethal and devastating genetic diseases such as Tay-Sachs, Lesch-Nyhan, and Huntington disease seems, at first glance, to be a good thing. At the same time, the desirability of eliminating a gene or a genetic defect associated with disease is complicated by the fact that we do not always entirely understand the gene's function. Some genetic defects associated with disease may also have a protective function. For example, the trait that causes sickle cell anemia also provides carriers with resistance to malaria. But recognition of this fact is surely only a reason to be cautious about genetic intervention, not a reason for an absolute bar. Malaria is not a major health problem in the United States, whereas sickle cell anemia causes pain, disability, and early death for thousands of African Americans. As long as attempts to improve the genome are not based on prejudice or misinformation, the claim that it is intrinsically wrong deliberately to change our genes is hard to understand. After all, we try in a variety of ways to improve our bodies, our minds, our lives. Why not our genes?

Effects on Families and Family Relations

Cloning would have a decided impact on our ordinary ways of determining family relationships. For example, if a man was cloned, would he be the father or the brother of the infant who was created? Would his parents be the child's grandparents or parents? Some commentators find the disruption to ordinary lines of kinship itself a reason to ban cloning. Others worry about the psychological impact of cloning on family relations. For example, Arthur Caplan has suggested that if a woman cloned herself, her husband might be sexually attracted to the child when she reached adolescence, realizing that she is not actually his daughter, but rather his wife's delayed genetic twin.

However, Caplan's objection is extremely speculative. Most spouses of an identical twin feel no romantic attraction toward the other twin.[11] This being the case, the fact that the child is his wife's twin should not cause her husband to become sexually attracted.

Perhaps Caplan's point is not that the husband would be attracted to his wife's twin, but rather that an important bar to sexual attraction between the man and the young woman, genetic connection, would be missing if the woman cloned herself. Because the husband would not be, strictly speaking, the father of a girl cloned from her mother, he might be sexually attracted to the girl. However, we should ask whether it is biological connection that prevents men from having sexual impulses toward their daughters, or the social role of father? If absence of genetic connection makes sexual feelings more likely, there would be an increase in father-daughter incest when a sperm donor was used or the child was adopted. I have no proof that this is not the case, but I strongly suspect it is not. My guess is that adoptive fathers and men who become fathers through sperm donation are at least as committed to being good fathers as men who become fathers biologically, and so are no more likely to molest their daughters. If the absence of genetic connection in these cases does not interfere with appropriate father-daughter feelings, why should it in the special case of cloning?

Some critics view cloning as the logical next step in a movement toward the rejection of traditional family values. For example, commentator Brian Brown says, "Cloning represents another assault on the traditional family." A family consists of a husband, a wife, and children who are the product of sexual intercourse. That's the correct model and anything else represents an attack on the family. Of course, this isn't the only kind of family we have today. There are blended marriages with stepparents and stepchildren; there are single parents and homosexual couples raising children; some children are not the genetic offspring of their parents, but are adopted; some children are born of surrogate arrangements; and some are conceived using sperm or egg donation. There are lots of ways to have children and form a family that do not conform to the "Leave It to Beaver" model. But these nontraditional families are precisely what Brian Brown objects to. His point is that liberals who accept these variations have no rational ground for drawing the line at cloning because the very morality that would support such concerns has been undercut. In other words, if single-parent families or homosexual couples are morally permissible, why not cloning? Brown writes,

> The elite that dominates Western societies has . . . pushed to normalize a whole range of acts that were once considered outrageous. And without shock or shame there is little left to trigger moral outrage. No one expedited this corrosion more than the shock troops of the 1960s who normalized shameful acts like promiscuous sex and drug use by ostentatiously promoting them. Now these very baby boomers decry the resulting social ills—from high levels of illegitimacy to drug-driven crime—without acknowledging their contribution to them.[12]

But this unlikely connection of cloning with sex, drugs, and (no doubt) rock 'n' roll, is not an argument against cloning; it's just right-wing raving. In any event, it is far from clear why technologies that enable people to have children and families constitute an "attack" on family values. Just the opposite appears to be true. When people who cannot have children in the usual way go to great trouble and expense to have children to raise, isn't that an affirmation of family values?

Admittedly, cloning would be a very different way of making babies. It is radically different even from assisted reproductive medicine, which still requires genetic material from a male and female. For some people, precisely the fact that cloning could result in the birth of a child without sexual reproduction is an argument against it. This view was expressed by a theologian on a television program who said about cloning, "It's not reproduction; it's replication! It's what starfish do!" Biologists tell me that this is inaccurate about starfish, but even if it were true, why should that make it wrong? Perhaps what the critics are getting at is that cloning would violate another alleged right: the right to have two genetic parents.

The Right to Have Two Genetic Parents

John Cardinal O'Connor expresses this objection when he says, "By design, a clone technically has no human parents, hence creating a clone violates the dignity of human procreation, the conjugal union (marriage) and the right to be conceived and born within and from marriage. A clone is a *product* made, not a *person* begotten."[13] Leon Kass makes a related objection when he notes that cloning severs procreation from sex, love, and intimacy and therefore is "inherently dehumanizing."[14] Of course, this objection applies not only to cloning, but to all assisted reproductive technology (ART), in which a baby is made not through lovemaking, but in the cold and sterile environment of a laboratory.[15]

Sexual intercourse, which both expresses the parents' love for each other and creates a physical embodiment of their love in the child, is a much better way of making a baby than ART. But some people cannot make a baby by making love, no matter how intense and loving their physical connection. Why should it be inherently dehumanizing for them to use the only method available to them to have a baby together? Less than ideal is not wrong.

The Catholic Church condemns virtually all forms of assisted reproduction, for reasons similar to those given by Kass.[16] However, Cardinal O'Connor's objection to cloning may not rest solely on its severing procreation from the conjugal act. His point may be that cloning is specially wrong, in a way that the rest of ART is not: Because the cloned child would be the delayed genetic twin of the somatic cell donor, not his or her offspring, cloning creates a child with no parents. Not having parents is a misfortune, a harm to the child, and therefore something to be avoided.

Some writers reject virtually all claims that birth should be avoided based on harm to offspring if the child has no other way to get born.[17] The argument, first made by Derek Parfit,[18] is based on the idea that avoiding birth cannot benefit the child who would have been born under adverse conditions; it can only prevent the child's birth. Whatever hardships or disadvantages the child may experience from the adverse conditions, he or she is not harmed, on balance, as long as his or her life is worth living. Consider, for example, commercial surrogate motherhood. One of the arguments against this sort of contractual arrangement is the possible harmful psychological effect on the child, stemming from the fact that his biological mother agreed to relinquish him at birth and get paid for it. Whether being born to a surrogate would have this harmful effect or whether it could be avoided is a debated empirical matter. However, some commentators are not much interested in the actual effects of surrogacy on the children. For them, as long as the child's life is, on balance, worth living, then surrogacy cannot be criticized as harming the child. That particular child would not have been born, but for the surrogacy agreement. Avoiding surrogacy cannot improve the child's life, but only prevent his birth.

This argument has implications for the debate about human cloning. As Robertson notes, "The problem arises because, but for the technique in question, the cloned person would not exist. Banning the technique may prevent a child from being born into the circumstances of concern, but it does so, not by assuring that it is born in different circumstances, but by preventing it from being born at all."[19] The question, then, is whether not having two genetic parents would make one's life so awful that never being born would be preferable. But this seems unlikely to be the case. Better a life without parents than no life at all. Therefore, Robertson would reject Cardinal O'Connor's argument against cloning based on harm to the child.

This is a complex issue and no one has resolved it satisfactorily yet.[20] On one hand, it is difficult to see how banning a technology can be "for the sake of the child" when that child would not have existed but for the technology, and the child has a life worth living. On the other hand, to dismiss as irrelevant to social policy or individual planning possible harmful effects of a technology, simply because the children who will come into existence as a result will not wish they had never been born, seems both callous and irresponsible. The question is how to explain the nature of the wrong. One possibility, suggested by Feinberg, is to explain the wrongness in terms of "wantonly introducing a certain evil into the world, not for harming, or for violating the rights of a person."[21] Yet this explanation seems unsatisfactory because it leaves out entirely the people who are most affected by the irresponsible behavior: the children who will lead seriously diminished lives.[22]

Although they disagree about how to explain the wrong, most commentators agree that it would be wrong to allow the development of reproductive technologies likely to result in the birth of children with emotional or physical impairments.[23] And most would agree that not having parents qualifies as a harm. Therefore, if cloning would create children without parents—orphans—that would be a reason for banning cloning.

However, it is far from clear that a child created by cloning would not have human parents. Consider this scenario in which cloning might be considered. A couple cannot have a child because of male factor infertility. Even ICSI, a technique whereby a single sperm is injected directly into the oocyte cytoplasm, fails to result in fertilization. The usual remedy is sperm donation. (Of course, using another man's sperm does not remedy the husband's infertility; it simply enables the wife to become pregnant.) However, sperm donation brings a third party (the sperm donor) into the marital relationship, which many couples would prefer to avoid. Indeed, it is precisely this feature of assisted reproduction that many of its critics find particularly objectionable. As Leon Kass expresses the point, "Clarity about who your parents are, clarity in the lines of generation, clarity about whose is whose, are the indispensable foundations of a sound family life. . . . Clarity about your origins is crucial for self-identity, itself important for self-respect."[24] For these reasons, then, the couple rejects sperm donation. However, they still would like a child biologically related to them, and the wife wants to have the experience of pregnancy and birth. SCNT cloning might make this possible. A somatic cell would be taken from the husband to obtain his DNA, which would then be put into an enucleated egg cell from the wife. The resulting embryo would be replaced in the wife's uterus for gestation. When they take that baby home from the hospital, it defies common sense to suggest that the child lacks human parents. Surely the woman who carried, gave birth to, and plans to rear her is her mother. Moreover, her husband has at least as much claim to be the baby's father as he would if they had used a sperm donor—perhaps a greater claim because there is a genetic connection between the husband and the child.

The Wisdom of Repugnance

Let me end with the objection raised by Leon Kass, who has called for an international legal ban on research on cloning humans.[25] Kass regards the possibility of human cloning as one more example of how we have been led astray by technology. It reflects our narcissism, obsession with control, and loss of the capacity for awe at the mysteries of nature and life. In response, I suggest that although cloning *might* express narcissism, it doesn't necessarily. In the situation I've just discussed, the charge of narcissism does not apply because the motive is the same as in other ARTs, indeed in ordinary reproduction: to enable the couple to have a baby biologically related to them. If this is narcissistic, then so is having children generally. As for obsession with control or loss of awe at the mysteries of nature and life, the same objection could be made to any medical advance.

Many people find the idea of cloning simply viscerally repugnant. Kass suggests that this repugnance, though not an argument, is "the emotional expression of deep wisdom, beyond reason's power fully to articulate it."[26] He suggests that we experience such repugnance at incest, bestiality, or cannibalism, even though we may not be able to give completely rational explanations of what is morally wrong with these practices. Yet we would be less than fully human, Kass maintains, if we were not repulsed by them. "Shallow are the souls that have forgotten how to shudder."[27]

The remark is stirring, but it avoids the central issue: At what should we shudder? We hardly need to be reminded that people have shuddered at all sorts of things, from blacks swimming in white swimming pools to children in daycare.

Shuddering may simply be a response to what is unfamiliar. As Sen-

ator Bill Frist, himself a former heart surgeon, pointed out in his testimony before NBAC, when the first heart transplant was performed in
1965, it too was considered by many to be wrong. The thought of transplanting the heart from a dead person into a living person's body made
a lot of people shudder. Now it's an accepted part of modern medicine.
The moral, it seems to me, is that we need to articulate the reasons for
objecting to cloning or any other new technology, and not rest content
with visceral responses. Until we do this, we have no basis for regarding the shuddering response as an expression of "deep wisdom." It
could simply be deep prejudice.

Conclusion

The acceptability of any potential reproductive technology depends
on the intentions and motivations with which it is done, as well as the
harms or benefits likely to result. These harms need not be strictly
observable or empirical. They might be intangible, symbolic, or spiri-
tual. But they do have to be articulated and made persuasive to people
from a wide variety of moral and philosophical beliefs. Only then will
we be able to continue profitably the discussion about the ethical implications of human cloning.

NOTES

Some of the material in this chapter is taken from my article "The NBAC Report
on Cloning Human Beings: What It Did—and Did Not—Do," *Jurimetrics* 38 (Fall
1997): 39–46.

1. Letter from William J. Clinton, President of the United States, to Harold Sha-
piro, Chair, National Bioethics Advisory Commission, Feb. 24, 1997, reprinted in
*Cloning Human Beings: Report and Recommendations of the National Bioethics
Advisory Commission,* 1997 (hereinafter *NBAC Report*).

2. "Will Cloning Beget Disaster?" *Wall Street Journal,* May 2, 1997, p. A14.

3. Robert Pollack, "Beyond Cloning," *New York Times* (op-ed), Nov. 17, 1993,
p. A27.

4. Joel Feinberg, "The Child's Right to an Open Future," *Freedom and Fulfillment* (Princeton, N.J.: Princeton University Press, 1992), 76–97.

5. Hans Jonas, *Philosophical Essays: From Ancient Creed to Technological Man* (En-

glewood Cliffs, N.J.: Prentice Hall, 1974), cited in Dan W. Brock, "Cloning Human Beings: An Assessment of the Ethical Issues Pro and Con," a paper commissioned by the NBAC and reprinted in *Ethical Issues in Modern Medicine*, 5th ed., ed. John Arras and Bonnie Steinbock (Mountain View, Calif.: Mayfield, 1998), 484–96.

6. *NBAC Report*, 32.

7. Gina Kolata, *Clone: The Road to Dolly and the Path Ahead* (New York: William T. Morrow, 1998), 38.

8. For a thorough treatment of the concept of exploitation, see Alan Wertheimer, *Exploitation* (Princeton, N.J.: Princeton University Press, 1996).

9. Martin S. Pernick, *The Black Stork: Eugenics and the Death of "Defective" Babies in American Medicine and Motion Pictures since 1915* (New York: Oxford University Press, 1996), 55.

10. Ibid., 95.

11. Steven Pinker, *How the Mind Works* (New York: W. W. Norton, 1997), 117.

12. Brian A. Brown, "Cloning: Where's the Outrage?" *Wall Street Journal*, Feb. 19, 1998, p. A22.

13. "Will Cloning Beget Disaster?"

14. Leon R. Kass, "The Wisdom of Repugnance," *New Republic*, June 2, 1997, p. 22.

15. See, for example, Paul Lauritzen, *Pursuing Parenthood: Ethical Issues in Assisted Reproduction* (Bloomington: Indiana University Press, 1993).

16. Vatican, Congregation for the Doctrine of the Faith, "Instruction on Respect for Human Life in Its Origin and on the Dignity of Procreation," *Origins* 16:40 (Mar. 19, 1987), reprinted in *Ethical Issues in Modern Medicine*, ed. Arras and Steinbock, 425–34.

17. See, for example, John Robertson, *Children of Choice: Freedom and the New Reproductive Technologies* (Princeton, N.J.: Princeton University Press, 1994), 75–76.

18. Derek Parfit, "On Doing the Best for Our Children," in *Ethics and Population*, ed. Michael D. Bayles (Cambridge, Mass.: Schenkman, 1976), 100–115.

19. John Robertson, "Liberty, Identity, and Human Cloning," *Texas Law Review* 76:6 (May 1998): 1371–456, quote on 1405.

20. For a persuasive if not entirely satisfactory account, see Dan W. Brock, "The Non-Identity Problem and Genetic Harms: The Case of Wrongful Handicaps," *Bioethics* 9 (1995), reprinted in *Ethical Issues in Modern Medicine*, ed. Arras and Steinbock, 397–401.

21. Joel Feinberg, "Wrongful Life and the Counterfactual Element in Harming," *Social Philosophy and Policy* 4 (1987), reprinted in Feinberg, *Freedom and Fulfillment*, 3–36.

22. Bonnie Steinbock and Ron McClamrock, "When Is Birth Unfair to the Child?" *Hastings Center Report* 24 (1994), reprinted in *Ethical Issues in Modern Medicine*, ed. Arras and Steinbock, 338–97.

23. Robertson may be an exception to this consensus. See *Children of Choice*, pp. 122 and 169, where he suggests that potential harm to offspring would not justify state-imposed bans except in "wrongful life" situations.

24. Leon Kass, "'Making Babies' Revisited," *Public Interest* 54 (Winter 1979), reprinted in *Ethical Issues in Modern Medicine*, 4th ed., ed. John Arras and Bonnie Steinbock (Mountain View, Calif.: Mayfield, 1995), 430–35, quote on 431.

25. Kass, "Wisdom of Repugnance," 22.

26. Ibid., 20.

27. Ibid.

JORGE L. A. GARCIA

5 Human Cloning: Never and Why Not

In one of Hegel's rare memorable passages, he remarks that the Owl of Minerva takes flight to paint its gray on gray at the end of day. He seems to have meant two things: that philosophy does little more than give intellectual expression to the spirit of the times and that it does even that rather late, as the *Zeitgeist* is itself changing. Whatever their truth as general claims about philosophy, they certainly capture the discipline of bioethics. Practices that once outraged the common sensibility are now all the rage. Bioethicists, true to Hegel's vision, have entered the stage wringing their hands over some new practice, but quickly changed their tune as social attitudes changed from hesitation and disapproval to cheery contentment. Indeed, as the first wave of medically and theologically trained writers on medical ethics has been replaced by today's crop of lawyers, policy specialists, analytic philosophers, and those who revel in the neologism *ethicists,* they have become so adept at this that they have gotten ahead of the curve of attitudinal shift. That is not to say that they actually cause change, but they have removed an important cautionary voice, a brake against brash and sweeping transformations. In the past, intellectuals played an important cultural role in cautioning against haste, calling for reflection, reminding of past troubles, articulating traditional cultural commitments and a sense of continuity with forebears, and so on. In contrast, today's secularized clerisy of ethics intellectuals are among the most vocal in assuaging any lingering moral doubts about the new agenda pushed by researchers and the increasingly consumer-driven, market-modeled medical industries. A December 1997 *New York Times* headline caught this phenomenon nicely in the area that concerns us here: "On Cloning Humans, 'Never' Turns Swiftly into 'Why Not'?" The story notes that after the initial near-unanimous outcry against cloning humans that immediately followed the announcement of the Dolly experiment's success, "scientists have become sanguine about the notion of . . . cloning a human being."[1] Bioethicists' uncharacteristically negative initial reaction to the renewed talk of cloning humans is now a cause of some embarrassment and, barely a year since Dr. Wilmut's announcement and less than a year after the National Bioethics Advisory Commission (NBAC) report, we were already in the midst of a full-scale moral reconsideration.

This is not a bad thing. Medical ethics probably suffers from too little reconsideration, not too much. Indeed, I think that one of the problems in some of the new literature on cloning is precisely that it treats as definitively settled moral questions about the status of the embryo and so-called preembryo, the moral legitimacy of abortion and in vitro fertilization done for more or less any reason, and other matters where some consensus may or may not be emerging among secular elites, but where nothing has really been proven morally, even if there is such a thing as moral proof. There is no reason, then, to decry the raising of the question the *Times* article heralds: Why not? However, the question should not be treated as an impatient challenge to put up or shut up, lest some new medical agendum be delayed. The philosophical approach is to treat the question as an inquiry into the reasons for which human cloning might be morally objectionable.[2] This is the spirit in which I will treat it here. Where the *Times* article notes a shift in attitudes from "never" to "why not"? the attitudes behind the two utterances are not opposed in principle. We can deny that cloning people is ever morally permissible and also inquire into what makes that true. Hence my subtitle, "Never *and* Why Not."

A Bioethicist's View of Human Cloning

Gregory Pence's *Who's Afraid of Human Cloning?* (1998) is one of the first book-length treatments by a philosophically trained bioethicist since the announcement of the Dolly experiment's success, which defends human cloning as ethical. For that reason, and the fact that it is being marketed to a mass audience as a general-interest paperback on current affairs and science, it warrants attention.

You might have thought that even if the initial reaction against human cloning was inadequately thought through, there are serious problems about the practice. Yet as Pence makes clear in his very title, this is not his view. Hesitation about its licitness is just a matter of fear, not reason. He warns us against "fear of a change, fear of changing human nature, fear of humans having more choice and control." This is just a struggle between the "fatalistic view . . . that everything is changing too fast" and those who distrust people, on one side, and "voluntarists," those who believe "we have the wisdom to use new knowledge to help people" and are "more optimistic," on the other.[3] Pence continues, "There is nothing about change itself," we discover, "that is bad." With this hard-won insight, he thinks, we can "take a more assertive stance toward the future of humanity."[4] One might think that the end of humanity's most destructive and barbarous century calls for more caution than Pence's none-too-searching question, "So why not trust humans rather than fear them?"[5] Reflect for a moment on what Arendt called "the banality of evil." Recall what nice, ordinary people did or let happen in Germany, or Alabama, or Siberia, or Soweto, or Tibet, in just the last few generations. Then follow the procedure Pence recommends in considering whether we might use the new technologies in ways that harm people: "Go to your local neighborhood meeting, Parents-Teacher Association night, or Kiwanis Club and ask yourself: are all those people the kind who have bad motives?"[6] Those with sufficient self-knowledge may not reach the answer Pence wants.

People always worry about new medical practices, Pence thinks, and the facts prove them wrong. After all, he says, "Physician-assisted dying for competent, terminal adults in Holland was predicted to turn that peaceful country into an ethical hell, but the practice has been going on twenty-five years with hardly any bad results."[7] Justice Souter, whose concurring opinion in the 1997 assisted-suicide cases legal commentators saw as almost inviting opportunity to find a more limited constitutionally protected right, seems to have held back in these cases largely because of the widespread abuses of the rules putatively governing physician-assisted suicide in the Netherlands.[8] Does not involuntary euthanasia, unreported and unpunished, count as a bad result?

Despite those concerned about dangers of widespread human cloning, we are not to worry: "For every high-minded couple who produced a superior child by NST [nuclear somatic transfer, a type of human cloning] there would be a Brazilian couple who produced nine children by normal sex."[9] Even within the often openly eugenicist discourse of many proponents of human cloning, this explicit contrast of the high-minded and superior on the one side and the Brazilian on the other is shocking. But it is presumably acceptable to speak of Latin American reproductive customs with open contempt because these people are likely to be Christians, especially Catholics or evangelicals. Those who are concerned about abortion and respect for human embryos are considered foolish extremists. Richard Lewontin has questioned the conspicuous absence of testimony before the president's commission from Christian fundamentalists.[10] For Pence, however, too much was heard even from mainline Protestant and Jewish thought, too much from religious people. Even government regulation of cloning and other techniques of artificial reproduction is to be avoided because government is too easily pressured by "extreme religious groups."[11] He does not seem to have in mind those whose views, like those of his intellectual hero, Joseph Fletcher, are extreme in their enthusiasm for new manipulations of and interventions in the beginning and end of life.

Lutheran theologian Gilbert Meilander worries that cloning might not comport with Genesis's picture of human beings divinely enjoined to sustain human life through procreation. According to Pence, "The problem with Protestants justifying their views on biblical passages is that they only go there to justify what they already believe, not to find guidance."[12] It seems rather harsh to accuse a believer of abusing his or her own scriptures, seeking in them not the word of the God he or she thinks therein revealed but only endorsement of existing prejudices. No fair-minded person would make such a charge against a person, let alone a whole sect, without first entertaining the possibility that the thinker might indeed have sought and received guidance from the passage, might indeed have read and reflected on it many times. Meilander holds that a child is a gift from God and that we should strive so to see it, a striving he thinks cloning repudiates and makes more difficult.

Pence is in a hurry and Meilander's "gift" talk threatens to slow things down. All this fretting about God and ethics is "holding hostage important medical research," after all.[13] That Pence cannot abide. "When are we allowed to choose to have better babies?" he asks. "Never? When are we allowed to say to the Giver of the gift, 'Gee, couldn't you do any better than that?'"[14] "Better babies," "superior children" for the "high-minded" through cloning? Or Latinos rutting away—beneath garish images of Jesus on the bedroom wall, no doubt—turning out their litters of human inferiors? The contrast latent in Pence's imagery is now manifest. We should turn to his more thematic discussion of the moral case against cloning. However, before we get to that, we must pause to clear up some confusions Pence introduces about moral reasoning.

Ethical Thinking

I have said that my interest here is to affirm the view that human cloning is not permissible morally (my "never") and to begin exploring some reasons for which it is not (my "why not"). If it is wrong, it is wrong for reasons. Notice, however, that this does not entail that in order for someone to know (or justifiably to believe) that it is wrong, he or she must first know why it is. It is an ontological point that has nothing to do with moral epistemology. Pence claims, for example, that "philosophers and bioethicists are very suspicious about 'knowing what you want to do' [in condemning something morally], but not knowing 'why' it is morally wrong." And plainly, he thinks their suspicions are right. In his view, "if the balance of reasons favors one side over another, we know that the right side is the one with the better reasons."[15] This may be right, but it is unwarranted. The reason for my doubt is precisely that this view does not seem to permit us to say that a position may be right but is unwarranted for us to assert at a certain time.

This principle would be nonsense as a general epistemological claim. If I claimed never to know what color or figural qualities a thing had until I knew why it had them, I would merely be deceiving myself. What is supposed to make the moral case so radically different? As in many other matters, we come to moral knowledge through various combinations of perception, testimony, inference, reflection, analysis, and empirical investigation. Of course, when I know that something is wrong, I often (but need not) also know some respect in which it is wrong. Still, it hardly follows that the "balance of reasons" must always favor the position that is, in fact, correct. Certainly, it need not if the reasons intended are merely the ones so far presented at a certain point in the discussion. Nor need the correct side even be favored by the balance of reasons available for our inspection. Maybe we just do not know yet what makes the thing wrong, as we do not know what grounds or causes many of its other qualities. Of course, there are forms of antirealism according to which saying something is wrong is just saying how the discussion of it is proceeding. And there are forms of constructivism according to which what is wrong is made wrong by a process of moral deliberation. I doubt any such metaethical theory is correct, but even if one proves true, that hardly warrants confidence that the correct moral position is always the one supported by the balance of reasons. So, as in other areas of inquiry, even if the moral arguments against human cloning were unpersuasive, weaker than those on the other side, that would not entail that the "right" view is that cloning is not wrong.

This recalls Peter Singer's skeptical approach to so-called moral intuitions. Singer asks why we should not distrust our intuitive moral judgments about particular cases as "derive[d] from discarded religious systems, from warped views of sex and bodily functions, or from customs necessary for the survival of the group in social and economic circumstances that now lie in the distant past? In which case," he continues, "it would be best to forget all about our particular moral judgments, and start again from as near as we can get to self-evident moral axioms."[16]

I cannot pursue the issue of moral epistemology here. Permit me just to observe that what is required is to show that the discarded religious systems are false; that, whatever their truth or falsity, these systems' moral views did not capture important truths about human beings and their needs, service to which may explain the endurance of those moral intuitions; that warped views of sex are more likely to be found in traditional views than in more modern ones; and that we are likely to come closer to self-evidence at the level of general principles than we are at the level of judgments about particular forms of behavior, such as human cloning. Indeed, Mill himself conceded that we are more certain about particular judgments we make about this lie or that assault than we are about such generalizations as the utilitarians' own happiness principle.[17] My own view is that the lesson to be learned from thinking about intuitions—general and particular, old and new—is that we should be distrustful of the least reliable of intuitive judgments, that is, those that have arisen recently to allow us to feel all right about ourselves as we engage in practices long recognized as perverse. However, that does much to undercut a line of reasoning popular among the fans of human cloning and other new medical practices. For example, they argue that cloning for sex selection, to tailor children to parents' (or others') design specifications, or as a source of tissue donations is not wrong because something similar is sometimes done using in vitro fertilization (IVF), where (they say) it does not elicit the horror it used to.[18] That frequency has eroded the sense of moral horror some people feel over such practices does not mean that they now pass some test of acceptability before respectable intuitions. I should say the same about the view that we have somehow generally come to know that human life does not exist in utero or, if it does, that it deserves no protection there—that we have a right to decide exactly how and when to die. (What of the man who chooses to go out in mid-orgy with the Spice Girls in the Super Bowl half-time show?) Where is our distrust of received moral opinion when we need it?

Returning to Pence, let us examine another claim about moral thinking and theory. He claims that Mill's famous harm principle, which holds that state prohibitions on liberty are permissible only when the prohibited behavior harms someone, "does not merely champion an area of personal life free from governmental interference, but also an area free from moral criticism."[19] It is difficult to see how this could this be right as an interpretation of Mill's principle, but what is more important is that it is difficult to see how it could be a correct moral principle.[20] If a range of my actions is free from *all* moral criticism, even my own, then how can I undertake moral reform by acknowledging my own past wrongdoing in that area and seeking to avoid such behavior in the future? Is that area of my conduct to be free only from other people's moral criticism? Then what room would there be for me to seek your moral guidance in a matter of my private life? Even if the state should not intervene, can it really be correct that there is nothing morally objectionable in my conducting my private affairs from racial or ethnic or gender or religious prejudice? Or is it that such conduct is wrong, but that nobody has any business telling me so? If so, then how do I learn to reform morally? And what becomes of freedom of speech in this new gag-ruled version of "liberalism"?

At this point, I turn to consider some of the principal moral objections raised against human cloning. My aim in the next section is not to develop any of these arguments into decisive proofs of the immorality of human cloning but merely to point out difficulties in some efforts to counter them.

Some Reasons *against* Human Cloning

Certainly, there is good reason to find the prospect of human cloning troubling. It appears in several ways to endanger society and those involved as donors or in gestation. It plainly poses a threat to the dignity and equality of women when, by plan, their childbearing loses its normal and proper origin in an act of spousal love. Pence realizes this possibility but poses no serious response. Instead, at this point he invokes his unusual interpretation of Mill's harm principle. Beyond that, he simply affirms that "women fearing increased sexism from the introduction of NST have a knock-down argument to any sexist fantasy about reproducti[ve exploitation] . . . they can simply refuse to get pregnant, refuse to stay pregnant, or refuse to gestate a fetus any more."[21] This comment misses the point of the objection several times over. The point is not about the consequences of desexed reproduction (i.e., whether it will increase sexism). Rather, it is about whether reproduction by human cloning already treats the gestating mother in a demeaning way.[22] In any case, it is no response to this concern to say that women can escape the degradation. For one thing, such ways out as sacrificing her child before its birth are already tragic. For another, a degradation eventually escaped is still a degradation and therefore something that should not be tolerated in the first place.

Similarly, there is good reason to worry that human cloning as it becomes widespread even as an available option depreciates and de-natures both sexual relations and reproduction by making the former merely one alternative among many for the latter. Consider the view of the sexual that Alan Goldman calls "plain sex." This view under-stands sexual activity in terms of sexual desire, itself conceived simply as desire for tactile bodily contact and its pleasures.[23] It clearly fails to capture the sexual. It does not even successfully differentiate sexual activity from a vengeful desire to poke somebody. Understanding sex in terms of sexual desire gets things backwards. It completely misses the sexual because the very term and its cognates enter our vocabulary in differentiating groups, organs, and activities defined by their role in a certain mode of reproduction. Not all sex does or should result in reproduction, of course, but the idea that we can conceptualize the realm of the sexual without mention of reproduction is one of those ideas it seems only a modern intellectual could have.[24] Others would know better.

Again, there is ground for concern that, at a time when it is con-ceded all around that family life is strained, difficult, and damaging especially to children, cloning muddies the concepts of family and parenthood. This is especially likely in some of the bizarre scenarios where, for example, a mother bears the clone of her own grandfather or herself. Lewontin claims each clone will have two parents, just like everyone else, apparently meaning the male and female whose chro-mosomes joined to shape the principal gene donor's genome.[25] Another writer suggests that "a clone may have four 'genetic' parents" plus two (or more) additional mothers.[26] What matters is that the mother may bear (and thus be gestational mother of) someone whose genetic parents (in Lewontin's sense of "parent") are her own great-grandparents. In another, she is gestational mother of someone whose genetic parents are her own. In still other scenarios, identical twins are born years, even decades, apart. What sense can we make of generations in such a family? Indeed, in what sense is it family when the relationships that constitute it no longer match those constitutive of family life? Some people are sanguine that the family can easily be "reconceived" or "revisioned." More sober minds will want to proceed with caution here with what Aristotle considered the fundamental unit of society. It is already broken in our culture, and there is every reason to suppose that cloning would only make it harder to fix. Strangely, although Pence touches on worries raised about the family here and there, he offers no sustained discussion of the impact of cloning on family. Instead, he brands such concern "hypocritical" on the grounds that there are other, more immediate steps we could take to protect families and children without bothering about cloning.[27] This ad hominem plainly does not rebut, or even address, the charge that human cloning could greatly exacerbate an already dangerously unstable social situation.

Human cloning may thus deprive the clone of real parents. She may have many quasiparents, but one ground for worry is that none may be tied to her in the role of protector that a child's parents traditionally occupy. This danger is aggravated to the extent that the clone's parents may be more likely than those of other children to have produced her merely as a means to their own ends (e.g., providing tissue for donation to other children) and to treat her accordingly.

There are other grounds for legitimate concern about particular forms of human cloning, but I will not pursue them here. Rather, I want to make a few remarks about one of the more serious objections to human cloning as intrinsically and decisively wrong. That is the claim that it wrongs the person cloned by degrading him. It strikes me as so transparently demeaning to a human being to make him a product of technological manufacture that it is difficult to understand why some people claim not to see it. This is *not* the way we have ever treated human beings; it *is* the way we have always treated the subhuman things we regard as wholly subject to our will. Thus, in cloning a human person is treated in a way otherwise reserved only for subhuman beings. It is hard to know a better definition of degrading or depreciating. Consider a religious perspective. For half a millennium, Trinitarians have praised God the Son as equal to the Father precisely as "begotten, not made." The clone, however, is made, not begotten.[28] Even some advocates of cloning consider it replication, not reproduction. It is hard to see equal treatment, much less acknowledgment of human equality, when one person is planned and designed by another and then manufactured to the latter's specifications. Of course, some people twist IVF and even sexual procreation in these directions. That shows not that these new perversions are morally unproblematic, but that they should be avoided and condemned everywhere and that forms of reproduction that facilitate or encourage them have a heavy moral presumption against them.

Nevertheless, Ruth Macklin told the NBAC, "If objectors to cloning can identify no greater harm than a supposed affront to the dignity of the human species, that is a flimsy basis on which to erect barriers to scientific research and its applications."[29] The report does not reproduce the context of her remarks, but it is important to observe that this quotation is not an argument but merely an assertion of her value ordering. I argue that conducting and applying (supposedly) scientific research is a pretty flimsy excuse for affronting human dignity. Of course, the person produced by cloning would not have existed but for this degradation. Some argue that this shows the act was not a net harm to her.[30] Even if that is correct, it does not suffice to show it is not a sufficient offense against her to render the act impermissible. After all, harm matters morally only insofar as it is a way of wronging someone. If harm is so narrowly defined that degrading someone is not harming her, then that only means that there are other ways of wronging people. So failure to harm does not entail failure to wrong. None of this means that the cloned person is subhuman, unequal, a thing to be used rather than a person to be respected. Rather, the argument presupposes just the opposite. That is why cloning is a degradation.[31]

In any case, pace Professor Macklin and others, scientific research is important, but we can do without it, as we did for most of our history. In contrast, it is doubtful that there is any secure foundation for human rights except in the inherent dignity of the individual. Thus, the Preamble to the 1948 *Universal Declaration of Human Rights* begins, "Whereas recognition of the inherent dignity and of the equal and inalienable rights of all members of the human family is the foundation of freedom, justice, and peace in the world." The first article, similarly, begins, "All human beings are born free and equal in dignity and rights."[32] The nature, source, limits, preconditions, and normative requirements of dignity could be made clearer, of course, as most important moral concepts could. That is a large part of the work of analytical moral philosophy. However, any suggestion that until this work is completed we should banish this concept—or the related concepts of rights, respect, and deference—from our discussion of the morality of human cloning should be regarded as we would the suggestion that we banish from bioethical discussions such controversial and imprecise concepts as cause, benefit, harm, or health until their conceptual clarification has been completed. We should greet it with derision.

If, as I maintain, all human cloning is wrong as a degradation of the one cloned, it may still be that some special forms of cloning are worse for special reasons. Thus, cloning *for* sex selection, to create tissue donors, to make "better babies," to replicate oneself, and so on further demean the child to the extent that they value her simply for her use and characteristics rather than her nature. Cloning of human multiples is especially repugnant. Cloning from grandparents and other ancestors is odious for the harm such arrangements may do this culture's already unstable family relationships.[33] Likewise, research toward human cloning should be rejected as immoral insofar as it destroys human "preembryos," encourages degrading views of humans as mere means to organs, pursues the loathsome eugenic project of "improving humanity" by manufacturing Pence's "superior children," and so on.

This research is morally impermissible in part for the reason Paul Ramsey identified: It is performed without informed consent from those experimented upon.[34] Some want to dismiss Ramsey's objection on the grounds that it is absurd to demand consent from someone to the very procedures that may bring him into existence. Of course, that is right, as Ramsey presumably knew.[35] What is unclear is why anyone should think this proves that such consent is inessential. There is no contradiction in saying that consent is required for morally acceptable research and also that it cannot be secured in some proposed research. What follows from this is simply what Ramsey said: The proposed research is impermissible. A defender of cloning experiments may not like this conclusion but still must give some rebuttal to Ramsey's argument.

Some may think they can use the commonly accepted principle that "ought" implies "can" to rebut Ramsey. After all, if this principle is correct, and if informed consent to the experiment cannot be secured, then the experimenter cannot be accused of wrongdoing for failing to secure it. Again, this is correct, but it does not effectively rebut Ramsey. For Ramsey's claim is not that the experimenter ought to (and, if the principle is correct, therefore can) secure consent. Rather, it is that the experimenter ought not to perform the experiment without consent. And the experimenter surely can refrain from performing the experiment without consent. He or she can abandon the experiment.

The Case *for* Human Cloning

I think the case for human cloning is rather weak. There is little to show that cloning would do much to protect rights, alleviate injustice, avoid treachery, promote virtue, or thwart vice. I suggest that one effort to vindicate the justice of human cloning fails rather badly. However, what matters is that even if it succeeded in demonstrating that position on the question of morally acceptable public policy, it could still be that human cloning itself is morally wrong—always, inherently, and indefeasibly. In short, the morality of public policy here, as elsewhere, underdetermines the central moral issue of whether the practice itself is morally permissible.

Some defend human cloning on the grounds that it could help prevent genetic disease.[36] Of course, this would be good. However, until we have some evidence of the likelihood that it really help and, more-

over, help in ways that could not otherwise be realized (or not otherwise realized without great sacrifice), this reason is, to borrow Macklin's term, "flimsy."[37]

The same holds for the defense of human cloning as an aid to those
afflicted with infertility.[38] How likely is it to help? How much? In what
ways? What are the prospects for alternative approaches? Moreover, we
should remember that although infertility is a genuine health dysfunc-
tion, there are already *many* legally permitted, and some morally permissible, ways of compensating to a greater or lesser extent (e.g., adoption, social volunteering, and assisted reproductive techniques). Insofar
as human cloning is proposed simply to assuage those unwilling or
unable to find such alternatives reasonable accommodations, the case
for it is still weaker. There is in general no compelling moral reason to
make sure everyone gets what she or he wants. Sometimes the proper
approach to dissatisfaction is to change one's desires, as the Stoics knew.
It is a lesson our culture needs to relearn, not least in these matters.

Eugenicist fans of human cloning think it will improve the race.[39]
This is no reason at all, for the supposed improvement is moral retro-
gression, as its vicious rhetoric of "superior children" should make
manifest. This merely displays an insulting and socially dangerous view
of illness and human limits. Healthier adults are not superior to others. The same holds for babies. The main reason some do not regard
talk of "better babies" as offensive is that some people, offensively, view
babies as functional items to be evaluated according to how well they
serve others' purposes, especially the purposes for which they were
made. This instrumental view of people is deeply wrongheaded and
ugly, yet it is the mentality that animates much of the push for human
cloning. There is a related point. Sometimes talk of preventing disease
is a smokescreen for eugenic improvements, as indicated, for example,
by Pence's enthusiasm for "changing our [human] natures."[40] What is
presented as a noble parental effort to avoid such illnesses as obesity
will sometimes result simply in parents refusing any child not up to
their standards of beauty.

Some also endorse cloning as a reproductive right.[41] I do not know
what the U.S. Supreme Court is willing to affirm as constitutional rights

these days. However, there is no good reason to see a *moral* right here. Although people plainly have some moral rights over their reproductive activities, talk of a right over how one reproduces is fanciful. Somebody might as well argue that a right to vote entails that Internet voting must be made available because some people would choose to vote that way.

Pence claims that homosexuals have been denied genetic connection to their children and endorses human cloning as a mode of redress.[42] Again, this is not serious for the same reason it would be unserious to demand such redress for celibates, avowed or adventitious.[43]

Theoretically, one of the most interesting arguments in support of the morality of human cloning is the appeal to John Rawls's theory of justice. Closely following Rawls, Pence reasons that in the original position, behind Rawls's "veil of ignorance," a rational contractor unaware of which generation she belonged to would choose for those in any generation to seek "the best genetic endowment" for their successors.[44] From this, he concludes that justice requires that society take steps to secure that optimal inheritance, including research and ultimately use of human cloning. Unfortunately, I think this argument is based on a misunderstanding of both Rawls's earlier and later understandings of his theory. Rawls's earlier version of his theory, in his book *A Theory of Justice,* makes it explicit that his theoretical apparatus is designed only to secure principles for ensuring that what he calls "the basic structure of society" meets criteria for "social justice." So understood, the apparatus of the original position is misapplied when used to derive conclusions about whether various practices are morally licit. Even if one accepts Rawls's theory as he first proposed it, the most that the defender of human cloning could show with it is that society should not interfere with human cloning, not that cloning itself is morally permissible. Of course, I doubt that Rawls's early theory, if itself correct, really shows even that. That the goal of eliminating genetic disease is justified does not suffice to show that such means as cloning are themselves permissible. Indeed, the deontological element in Rawls makes it more difficult to derive such conclusions about means from premises simply about ends.

The theory from Rawls's later book, *Political Liberalism,* is still more narrowly circumscribed as a theory simply of political justice for reaching collective political decisions in societies with certain kinds of history, commitments, projects, self-conceptions, and so on. Again, it contains no conclusions about the permissibility of such nonpolitical practices as human cloning.

On the whole, then, the case for human cloning as found in such works as Pence's is hardly compelling. For the most part, it does not deal in the graver moral realms of freeing people from injustice, ending vicious conduct, or attaining a deeper appreciation of what is valuable. Nor is it at all established that human cloning is likely to free real people from what any reasonable, objective observer would see as serious health deficits in someone's functioning as a human being. Rather, often the case largely reduces to the claim that human cloning may make some things go somewhat better for some people, largely by making things go more to their liking. In light, among other things, of the affront to human dignity and equality that human cloning appears to constitute, a much stronger kind of defense is needed to vindicate it morally.

Conclusion

In the end, human cloning merely presents us with a different face of what has rightly been called the "anti-life culture" that infects our medicine. It is another way of attacking human life, this time by degrading it rather than destroying it, by treating human life as something for us to bestow, and therefore of subordinate and only instrumental value. That other practices manifesting this mentality have won wide public acceptance in the last few decades does nothing to justify them, let alone human cloning.

NOTES

I am grateful to the audience at University of San Francisco's 1998 conference "Human Cloning: Science, Ethics, and Public Policy" and to Victoria Wiesner and W. David Solomon for bibliographic materials.

1. Gina Kolata, "On Cloning Humans, 'Never' Turns into 'Why Not,'" *New York Times,* Dec. 2, 1997, p. A1. See also George Johnson, "Ethical Fears Aside, Science Plunges On," *New York Times,* Dec. 7, 1997, p. 6.

2. A usage note: Throughout, I talk simply of "human cloning." This is not as clear as it could be because the term can be applied to many different things. On this, see Gregory Pence, *Who's Afraid of Human Cloning?* (Lanham, Md.: Rowman & Littlefield, 1998), 11. However, many of these differences make little moral difference and it is important to resist the urge to drift into technical obscurity and lose the resonance of the more familiar term. Pence and others now prefer the term *nuclear somatic transfer* (ibid., 49). I demur, concerned lest this move, like the insistence on the term *preembryo* and similar moves, form part of a strategy of obfuscation and euphemism. On the strategy, with special reference to the terms *preembryo* and *nuclear somatic transfer,* see Kolata, "On Cloning Humans," p. F4.

3. Pence, *Who's Afraid of Human Cloning?,* 123–25, 139, 165, and passim.

4. Ibid., 7.

5. Ibid., 65.

6. Ibid., 66.

7. Ibid., 70.

8. See Ronald Dworkin, "Assisted Suicide: What the Court Really Said," *New York Review of Books,* Sept. 25, 1997, pp. 40–44. On Holland's troubles, see Herbert Hendin, *Seduced by Death: Doctors, Patients, and the Dutch Cure* (New York: W. W. Norton, 1996).

9. Pence, *Who's Afraid of Human Cloning?,* 130.

10. Richard Lewontin, "Confusion over Cloning," *New York Review of Books,* Oct. 23, 1997, p. 23.

11. Pence, *Who's Afraid of Human Cloning?,* 35, 153.

12. Ibid., 80.

13. Ibid., 97.

14. Ibid., 81.

15. Ibid., 5, 6.

16. Peter Singer, "Sidgwick and Reflective Equilibrium," *Monist* 58 (1974): 516, quoted in Pence, *Who's Afraid of Human Cloning?,* 64.

17. Note, too, that whatever Mill may have thought, it is no longer plausible to maintain that utilitarianism can stand without support from our intuitions either in its consequentialist account of what actions are right, its aggregative and sum ranking account of what distributional schemes or states of affairs are better, or its account of the maximand. Indeed, the very decision to interpret its central principle as that of maximizing what is good rather than one of minimizing what is bad is itself usually decided on the basis of intuition. There is no obvious warrant for Singer's—and, by implication, Pence's—confidence that all these intuitions will be trustworthy and pristine whereas intuitions about the immorality of such practices as cloning are corrupted. For an introduction to some of the issues over utilitarians' competing understandings of happiness as pleasure or preference satis-

faction, negative utilitarianism, act utilitarianism versus rule and other forms of indirect utilitarianism, whether the happiness principle should be used by agents in practical deliberation or only by critics in retrospective assessment, and other related issues, see Geoffrey Scarre, *Utilitarianism* (London: Routledge, 1996).

18. See, for example, John A. Robertson, "The Question of Human Cloning," *Hastings Center Report*, Mar./Apr. 1994, p. 11. See also Richard McCormick's response, "Blastomere Separation: Some Concerns," ibid., 14–16.

19. Pence, *Who's Afraid of Human Cloning?*, 142.

20. As an interpretation of Mill, the problem is that it is not consistent with the utility principle to maintain that there are actions immune from moral assessment.

21. Pence, *Who's Afraid of Human Cloning?*, 145.

22. It is not fully clear that this makes it sexist because, although the gestating mother must be a woman, of course, even males involved in desexed reproduction and child-rearing will similarly be demeaned by their participation.

23. Pence, *Who's Afraid of Human Cloning?*, 79.

24. Or still worse, I suppose, a postmodernist one. As an example, see Michel Foucault's three-volume history of sexuality, *A History of Sexuality* (New York: Vintage, 1990).

25. I ignore the complication of mitochondrial genes.

26. Lewontin, "Confusion over Cloning," 21; Pence, *Who's Afraid of Human Cloning?*, 122–23.

27. Pence, *Who's Afraid of Human Cloning?*, 139–40.

28. See Albert S. Moraczewski "Cloning Testimony," *Ethics and Medics* 22 (May 1997): 3; Pontifical Academy for Life, "Human Cloning Is Immoral," *The Pope Speaks* 43 (Jan./Feb. 1998): 27–32.

29. National Bioethics Advisory Commission, *Cloning Human Beings* (Rockville, Md.: NBAC, 1997), 71.

30. See the discussion in ibid., 65–66.

31. Moraczewski, "Cloning Testimony," 3.

32. See, for example, the text reproduced in the periodical *First Things* 82 (Apr. 1998): 28–30. For a discussion, especially on the importance and ground of human dignity, see the Ramsey Colloquium, "On Human Rights," ibid., 18–22. See also Richard Doerflinger, *National Conference of Catholic Bishops' Statement at U.S. Capitol* (Washington, D.C.: NCCB, Jan. 29, 1998); Moraczewski, "Cloning Testimony"; and Pontifical Academy for Life, "Human Cloning Is Immoral."

33. It may be that cloning *for* male or female homosexuals, cloning the superannuated, and so on are also to be condemned for similar reasons.

34. See Paul Ramsey, *Fabricated Man* (New Haven, Conn.: Yale University Press, 1970). See also Pence, *Who's Afraid of Human Cloning?*, 52.

35. Notice, however, that it begs an important question about the moral status of the embryo (and "preembryo") to assume that there is no person at all involved in cases of experimentation on a human embryo. I am grateful to Al Howsepian for focussing my attention on this element in the dispute between Ramsey and

Pence. The common ground in such cases is that there is no person in a position to give or withhold consent.

36. Pence, *Who's Afraid of Human Cloning?*, 101–6.

37. At least here it can be said that human cloning would pursue this end in a less morally outrageous way, that is, by genetically healing those with genetic disease markers, as compared with IVF, where the effort is not to help any real person but to prevent conception of the diseased or to destroy those marked before they can be born.

38. Pence, *Who's Afraid of Human Cloning?*, 106ff.

39. Ibid., 166–70.

40. "A frequent corollary to the fatalist viewpoint is that human nature is not to be trusted with any new knowledge. Any attempt to change our natures [in this view] will produce dark consequences" (ibid., 124; see also 165). Note that Pence here talks as if gaining new knowledge and trying to change human nature were the same thing.

41. Ibid., 44, 45, 101.

42. Ibid., 114.

43. Moreover, it merely encourages irresponsibility to disconnect natural effects from causes in this way. In light of what we said about the family, it can be seen that bioengineering is partner of dangerous, unproven social experiments. Those who, like Pence, pride themselves on the empiricism of their approach need to attend more closely to the safety issues surrounding these social experiments.

44. Pence, *Who's Afraid of Human Cloning?*, 112–14.

PART 3

Cloning and
Public Policy

6 Cloning and the Ethics of Public Policy

Much of the discussion about cloning focuses on whether the act of cloning is right or wrong, according to some theory of morality. This is a worthy debate, and one that should help people to decide for themselves whether to take advantage of this technology, should it become available to them. But at the level of public policy, there is a different debate, namely, whether cloning should be forbidden or permitted. It is insufficient to argue that cloning is wrong and therefore should be forbidden. Many things are wrong but nonetheless are permitted, including bad manners, lying (except, for example, when under oath), and having a child when one is too young or impoverished to care for it as one should. The reasons for avoiding a government prohibition of such bad acts are many. For example, government action might be ineffective, overly intrusive, oppressive, or harmful to third parties. In other words, at times government prohibitions are a cure that is worse than the disease.

In the public testimony and submissions to the National Bioethics Advisory Commission (NBAC), three kinds of arguments were heard. Each represented a different kind of analysis that led to a conclusion that cloning is wrong. And each was linked to a particular set of policy concerns. In the end, NBAC found that all but one of the arguments for the wrongness of cloning failed to overcome the obstacles to enshrining that moral conclusion in law.

The first set of arguments focused on the motivations of those who might want to use cloning, a "relational" argument. By this analysis, cloning is neither intrinsically good nor intrinsically bad. Rather, when done for good reasons, it is ethically acceptable. When the motivations are bad, the act is bad as well.

By way of example, some people argued that if cloning were used to circumvent infertility, it is being pursued for a worthy reason and is not a bad act. Others suggested that having a child who is a genetic twin of a sibling in need of a bone marrow transplant is arguably good because it can save the life of one person without undue pain or suffering to the other. Here, too, the suggestion was made that in such cases cloning is not a bad act and need not be forbidden. On the other hand, these people argued, when cloning is used to satisfy one's ego or when it is used in conjunction with commercialized eugenics (e.g., by selling embryos cloned from "desirable" people at a price higher than that of "undesirable" people), then it is evil and should be outlawed.

In many ways, this approach accords better with moral intuition than many of the analytical arguments that follow. As a matter of public policy, however, it is just not feasible to implement. There is a history of miserable experiences when trying to create rules in which the government defined which acts are permitted and prohibited based on the motivations of the actors. For example, many people who support legal abortion are appalled at the notion that it could be used by someone who simply wants to select for the sex of a child. They find this inherently sexist or, at least, unacceptably gratuitous as a justification for abortion. Therefore, these people often want to prohibit abortion for this one particular reason while preserving all other reasons for allowing abortion.

But the states have found the implementation of this public policy unworkable. It requires an inquiry into the hidden psyches of people who propose to have an abortion, an inquiry that is inherently intrusive and subject to fraud and manipulation. Indeed, it hearkens back to the pre-Roe era in which the "liberal" states permitted abortion if a woman could persuade a panel of physicians that her reasons were adequate, a process that invited women to dissemble in order to obtain permission. In the end, the procedure was demeaning to the women and failed to achieve its purposes. One could predict a similar phenomenon should a relational ethic undergird a public policy that premises permission to use cloning on a sufficiently persuasive case being made to some appointed body of judges.

The second type of argument heard by NBAC was deontological, the "thou shalt not" school of reasoning. Simply put, according to this analysis cloning is wrong because it violates certain fundamental rules about the appropriate relationship between humans and nature (or humans and god), takes away the necessity of sexual intercourse, and confuses our common understandings of kinship and the separation of generations. These assertions were well illustrated by the moral repugnance arguments of Leon Kass. He, and others like him, argued persuasively that we should take moral intuition seriously, as it bespeaks a deep attachment to certain fundamental values that, however poorly articulated, nonetheless bind us together as a culture.

But whether these intuitions can be a sufficient basis for public policy is a separate question. To the extent that they are premised on explicitly religious convictions, they are insufficient because no one religious view can be imposed on others. The First Amendment to the Constitution guarantees that. Of course, if religious views can be supported by arguments that even nonbelievers can understand and accept, then they can become a basis of public policy. All that is needed is a widespread consensus.

But a consensus, backed up by a popular vote, can become something more than popular democracy at work. It can become oppressive if it stifles the preferences of a dissenting minority. This is tolerated for most situations, where a simple process of "majority rules" is accepted. It is not tolerated, however, when the interests being stifled are considered fundamental to our notions of liberty.

These so-called fundamental interests have been identified, albeit imperfectly, by the Supreme Court. They include the things specifically mentioned in the Bill of Rights, such as freedom of speech, assembly, worship, and association. They also include things so basic that no one would think they need to be listed in the Constitution, such as the right of heterosexual couples to marry and form a family. And they include the things that are implicit in the decisions of the Supreme Court by virtue of their close relationship to these aforementioned rights. The fundamental right to privacy falls into this category.

Thus, before one can accept a public policy based on a consensus that cloning is wrong and a popular vote to prohibit its use, one must ask whether such a prohibition will impinge on a fundamental right. Here, the waters become murky. Cloning is arguably just another form of reproduction. The Supreme Court decisions since the mid-twentieth century have carved out a large area of family life and reproductive decision making as beyond the appropriate scope of government authority, on the basis that these implicate fundamental rights. But there has never been a clear statement that there is a fundamental right to procreate, especially to procreate by means that require third-party assistance.

Further complicating matters is the question of whether cloning is a form of procreation, in either the biological or social sense. Although it surely is a form of making a new person, it is divorced from the sexual interaction that forms the emotional and social underpinnings of the experience that has led courts to claim a zone of privacy for American citizens. Thus, it is worth questioning whether the Supreme Court, were it to revisit those earlier cases in light of this new development, would premise its decisions on a right to make a new person or some other kind of right, such as the right to form intimate associations with others that may entail family formation.

If it is true, though, that cloning is at least arguably a form of exercising a fundamental right, then reasons for its prohibition must go beyond moral repugnance. They must be consequentialist, the third category of argumentation heard by NBAC. In other words, to abridge a fundamental right, one must show that there is a compelling state purpose to be achieved, such as the prevention of a probable and serious harm, and that the only way to achieve this purpose is to abridge the right. Thus, many people argued that cloning would harm children and society. Children, they argued, would be harmed by the excessive expectations brought to their births by the adults who believed that genetic duplication guarantees duplication of all physical and even psychological traits. Even though this is untrue, such expectations of the parents would cause the harm. In addition, they argued, the child might have false expectations based on the circumstances of his or her birth and thus lose the unexpected quality of an open future that is the norm for children. Others argued that a child conceived through cloning would suffer a form of genealogical bewilderment, as he or she might well have a rearing parent who was a biological sibling.

People also argued that cloning would be harmful to society because it would encourage a kind of commercialization, or at least commodification, of children. Images of mass marketing in cloned embryos featured heavily in testimony and newspaper letters to the editor. And the merging of images of cloning and images of soulless drones gave rise to widespread fears of whole populations being created for slave labor. Should all these things come to pass, they might well be harmful, and some might even meet the test for a compelling purpose. But these harms were both vague and uncertain, hardly enough to overcome the obstacles to abridging what might be a fundamental right.

In the end, NBAC identified only one kind of harm that met this test, and this was the possibility of physical side effects from the cloning procedure that could result in injury to the children. In the near total absence of animal data, and in the presence of suggestive hints of physical problems that might be associated with cloning (ranging from atypically high mutation rates to shortened life span), it seemed appropriate to limit use of the technologies. In keeping with a policy of limiting rights as little as possible, however, NBAC recommended that the prohibition be of five years' duration only, while animal data accumulated. Then a reassessment could take place.

NBAC did consider other ways to protect children from harms associated with premature use of cloning in humans. For example, it considered whether the threat of medical malpractice suits would be sufficient to dissuade careless use of the technology. If a person acted in a way that was unreasonably careless, then that person could be punished in the form of monetary damages. This result would send a signal to the rest of the community and serve as a deterrent for similar behavior. By retrospective punishment of an individual, society deters acting in the same way and changes the standards of care. Unfortunately, this is not a particularly satisfying solution. Judging somebody to be unreasonably careless is a difficult task. It is easy to imagine that driving significantly above the speed limit is unreasonable behavior. In a rapidly evolving field, however, there is no professional speed limit, no established point of reference. For this reason, among others, tort and malpractice law were unlikely candidates for effective regulation of human cloning.

Alternatively, NBAC considered a voluntary moratorium in which physicians and researchers in the United States would agree not to attempt human cloning until a certain period of time had elapsed to allow sufficient data of animal cloning to accumulate. NBAC commissioned a study on the history of voluntary moratorium in the area of genetic engineering. With a few exceptions, this method worked extremely well in terms of achieving self-restraint in the biological community. However, NBAC could not ignore the lack of federal oversight in the field of in vitro fertilization, which was an artifact of attempts by the federal government to divorce itself from abortion politics. An absence of federal funding has meant an absence of federal oversight.

Reproductive technologies are characterized by a confusion between research and therapy, experimental and standard care, all of which means that patients are poorly protected from overly zealous practitioners and from the exigencies of the marketplace. NBAC was unable to get assurances from the most relevant professional societies in the field that they would try to achieve a voluntary moratorium. The absence of such assurances from societies representing the very professionals who might be interested in offering the technology meant that a voluntary moratorium was not a realistic means for curbing premature and possibly unsafe experimentation.

At the end of the day, many people complained that NBAC had failed in its duty. By never issuing a stinging condemnation of cloning, by never endorsing any one of the many arguments claiming its inherent wrongness, NBAC had failed to grapple with the issues, they claimed. But that is not the case. NBAC is not a group of individuals speaking for themselves, nor a group of religious leaders interpreting a text. It is not the authoritative moral voice of America. Rather, it is a committee of people dedicated to offering advice on how to develop pubic policy that is ethically defensible. This means that the ethics of the policies as well the ethics of the underlying acts are at issue. For this reason, NBAC considered its job to be to integrate arguments about the ethics of cloning with an understanding of American political and legal culture.

Newsweek ran a piece many years ago, before the end of the Cold War, that recited a little ditty attempting to explain differences in national political cultures. It went something like this: In the United States, everything is allowed unless it is specifically prohibited; in East Germany, everything is prohibited unless it is specifically allowed; in the Soviet Union, everything is prohibited especially if it is allowed; and in Italy, everything is allowed especially if it is prohibited. Though casual and perhaps too cute, this ditty nonetheless captures some fundamental approaches to governance. NBAC took this advice to heart.

NOTE

This chapter previously appeared as "Cloning: Ethics and Public Policy," 27 *Hofstra Law Review* (1999), and is reprinted (with minor editing changes) with the permission of the Hofstra Law Review Association. An earlier version was presented at a conference on human cloning, April 3–4, 1998, at the University of San Francisco.

ANDREA L. BONNICKSEN

7   Crafting Cloning Policies

Although entrepreneurs who have announced their intention to clone humans may disagree, we are unlikely suddenly to read of the birth of a baby following somatic cell nuclear transfer (SCNT) or to see a clear line announcing a precloning and postcloning era.[1] Instead, we will see many steps moving us closer to human cloning by establishing the science of cloning and by reducing our reservations about cloning through the acceptance of related procedures that produce some medical benefit. This is not to say that cloning is inevitable; rather, the path to cloning will be filled with milestones that, depending on social mores, lend the occasion either to proceed deliberately or to draw lines on the road to cloning.

This chapter develops the theme that a technological incremental-ism is under way for cloning that lends itself to a policy incremental-ism. It argues that subtle developments in science and medicine will reduce public resistance to cloning and that banning one type of clon-ing now will not necessarily stop future developments. Moreover, a sub-dued public response to the somatic cell cloning of mice in 1998 and the growing strength of groups opposed to a cloning ban suggest that leg-islation to outlaw cloning permanently in the United States may not be forthcoming or that laws passed will be contested in court. This in turn underscores the need to develop contingent policies to guide cloning's gradual emergence and to draw lines where appropriate. The medical community is a potential first-tier builder of this policy that generally is overlooked in debates oriented to legislative action.

Incremental Steps to Cloning

Several developments point to the incremental and gradual onset of cloning technologies, which may or may not culminate in SCNT for humans, depending on whether lines are drawn midstream. First, researchers using animal models are making significant advances in cloning, with many more accomplishments on the way. Researchers have experimented successfully with cloning by nuclear transfer using embryo, fetal, and adult cells and across mammalian species in order to replicate prize stock, manipulate animals genetically for pharmacologic use, and pursue other scientific and commercial ends. They have cloned lambs and rhesus monkeys, among other species, using cells from early- or late-stage embryos and they have cloned lambs, calves, and mice using differentiated cells from animal fetuses[2] and from adult animals. In the same experiment that produced Dolly the cloned lamb from the mammary gland cells of an adult ewe in 1996, for example, three lambs were reported born from fetal cloning.[3] In mid-1998, researchers using the cumulus cells surrounding adult mouse eggs generated mouse clones and clones of those pups.[4]

Researchers have also combined cloning with the genetic manipulation of animals for pharmacologic and medical use. Faced with an inefficient method of injecting genes into embryo cells individually, which is slow because only a small number of embryos assimilate and express the injected gene, scientists have used cloning to produce many cells for gene splicing.[5] The successful pairing of cloning with transgenic animal research opens new horizons of cloning research.

Second, research into potential therapies in humans can be anticipated that will indirectly pave the way for human cloning. For example, egg cell nuclear transfer is an impractical technique in which nuclei from human egg cells would be transferred to enucleated donor eggs for fertilization and transfer. This would be done either to facilitate infertility treatment or to prevent the birth of infants with serious genetic disease.[6] In regard to infertility treatment, researchers have hypothesized that the fertilizability of the eggs of women over 40 is compromised because of the cell's aging cytoplasm. If this is true, then transplanting the patient's egg nucleus to the enucleated egg of a younger donor might aid fertilization.[7] It has also been hypothesized that egg freezing rarely is successful because the freeze/thaw harms the cytoplasm. If this is true, then transplanting the nucleus from a frozen/thawed egg to a fresh enucleated donor egg might permit successful egg freezing.[8]

In regard to the latter, egg cell nuclear transfer has also been proposed to circumvent mitochondrial disease for the offspring of women who want to conceive with their own gametes.[9] Mitochondrial DNA (mtDNA) reside in the cytoplasm of all cells except the spermatozoa; disease-linked defects in mtDNA are passed from a woman to her children because her cytoplasm becomes part of the developing embryo. If the nucleus from her egg were transplanted to the enucleated egg of a donor without the disease, the child would not inherit the disease but would still inherit the mother's nuclear DNA. Egg cell nuclear transfer is not cloning because no genome is replicated, but it would advance cloning by providing a less controversial setting for perfecting the nuclear transfer procedure, which underlies cloning.

Third, germline gene therapy (GLGT) is another area of hypothetical human clinical application that may indirectly pave the way for cloning. With GLGT, the nuclear DNA of gametes (eggs or sperm) or embryos would be manipulated to correct a disease gene; presumably, the intervention would be inherited by subsequent generations. Although anticipated with apprehension, GLGT has more recently gained advocates, including those who outline its possible underlying science.[10] Although GLGT is not cloning, it may be facilitated by embryo or other cloning, which would provide greater numbers of cells for gene splicing. Specialists in germline manipulations on animals are more enthusiastic about Polly, a transgenic cloned lamb, than Dolly, who was merely cloned.[11] The intersection of cloning and germline interventions in animal models suggests that if GLGT in humans gains legitimacy, cloning to replicate genomes to facilitate GLGT but not to generate individuals sharing a genome may be a step in the process.

Fourth, human embryo twinning to increase the number of embryos for transfer for a couple who produce only one or two embryos is a possibility.[12] Although negative public reaction greeted news of an early experimental study of human embryo twinning,[13] as time has passed views have moderated somewhat, and the National Advisory Board for Ethics in Reproduction (NABER) and Ethics Committee of the American Society for Reproductive Medicine both concluded that twinning is not necessarily unethical under controlled conditions.[14] The birth of a healthy baby following twinning to increase the number of embryos or to aid in preimplantation diagnosis probably would dissipate resistance to the idea of genome duplication, which would undercut the anticloning argument that each child has a right to genetic individuality.[15]

Current Policy Context

Policy efforts in cloning have been crisis driven, with periods of calm
p   ⸱⸱⸱⸱ted by bouts of activity, which make an ill fit with incremen-
⸒   ogical growth. The announcement of Dolly's birth in 1997
           ⸱d one perceived crisis whereby within weeks legislators in
         tates proposed more than sixteen anticloning bills, legislators
         ⸱d three bills to the U.S. Congress, and President Clinton called
         ⸱ratorium on human cloning, ordered that no federal funds be
         ⸱ human cloning research, drafted an anticloning bill for Con-
         ⸱nd charged the National Bioethics Advisory Board (NBAC) to
         rate on cloning and report to the public in three months.
         ⸱ese responses were largely symbolic. Federal policy already barred
         ing research involving human embryos, which would extend to
         ⸱ryos generated through cloning. President Clinton's proposed bill
         ⸱ lengthy and not carefully constructed, and no legislators sponsored
         His imposition of a three-month deadline on NBAC's deliberations
         as unreasonable in light of cloning's significance, complexity, and
         ⸱ypothetical status.

No tangible policy emerged in the United States after the first perceived crisis, and momentum lagged until a second crisis, this one more distinctly media driven, emerged in early December 1997 when Richard Seed, a physicist and entrepreneur without solid credentials in the field of assisted conception, announced his intention to clone humans.

This prompted a second set of policy responses. By this time nineteen states had introduced more than twenty-two cloning bills, and one state, California, had passed an anticloning law. The crisis atmosphere surfaced at the federal level when the Senate majority leader tried to move an anticloning bill directly to the Senate floor within forty-eight hours of its presentation and without hearings.[16]

The effort to push an anticloning bill through the U.S. Congress seriously miscalculated the strength of interest group politics, when more than seventy biotechnology, medical, and patient advocacy groups held press conferences and argued before Congress that efforts to discourage reproductive cloning made sense, but that a broadly worded ban would stymie medical research geared to treating diseases and conditions. Among the forty-five groups signing one letter of concern were the American Academy of Pediatrics, American Heart Association, Asthma and Allergy Foundation of America, and Cancer Research Foundation of America. In addition, the American Society of Cell Biology circulated a letter signed by twenty-seven Nobel laureates urging caution and a voluntary rather than legal moratorium.[17]

Recognizing the complexity of the issue, senators Dianne Feinstein and Ted Kennedy started a filibuster on the effort to bring the bill to immediate vote. Majority leader Trent Lott tried to close debate, but in testimony to the strength of biotechnology and medical and pharmaceutical interests, the effort to close debate was defeated in a forty-two to fifty-four vote.[18]

A number of bills had been introduced in Congress by late 1999 that would ban the practice of human SCNT, research into human SCNT, or the federal funding of cloning or cloning research. For example, Senate Bill 1602, sponsored by Democratic senators Feinstein and Kennedy, would prohibit any person, public or private, from transferring an embryo generated through SCNT to a woman's uterus for procreation, and it would bar the use of federal funds for this purpose.[19] Senate Bill 1601, sponsored by Republican senators Lott, Bond, and others, would prohibit any person, public or private, from using human SCNT technology. This would ban human SCNT research as well as the actual transfer of an embryo to a woman's uterus.[20] In the House of Representatives,

H.R. 922, which passed the House Committee on Science in 1997 but had not gone to the floor by late 1999, would prohibit the use of federal funds to support any research that "includes the use of human somatic cell nuclear transfer technology to produce an embryo."[21] Although these bills differ, they share a theme that human SCNT should not be used for procreation, nor should federal money be spent for this purpose. At present, efforts to ban cloning to procreate and cloning research are on hold at the federal level.

At the state level, legislators introduced more than twenty cloning bills in nineteen states. Three states—California, Michigan, and Rhode Island—have passed anticloning laws. Cloning bills in the states vary in many ways. If enacted, some would impose administrative penalties, whereas others would amend the criminal code. Some would make cloning a crime, and others also would make conspiracy to clone a crime. Some would exempt activities such as in vitro fertilization or twinning and preimplantation genetic diagnosis; others would include additional techniques in their reach, such as cross-species fertilization. The bills vary in how they define cloning, so it is not always clear what is being prohibited and why. Perhaps all variants of reproductive cloning should be prohibited or perhaps none or only some should be; the debate has not been so refined as to consider various forms of SCNT. At the federal level, only adult SCNT (the form used to generate Dolly) is targeted in the principal bills, but assorted definitions at the state level leave the legislative intentions unclear.

California's law is instructive because it is the first cloning law on the books in this country. Under California's law, "'Clone' means the practice of creating or attempting to create a human being by transferring the nucleus from a human cell from whatever source into a human egg cell from which the nucleus has been removed for the purpose of [initiating a pregnancy]."[22] By including cells "from whatever source," the law forbids embryo nuclear transfer and fetal and adult cell nuclear transfer in the absence of preliminary public discussion of anything beyond adult cloning. It also forbids egg cell nuclear transfer, which is not cloning at all because no genome is replicated.

California is perceived as an innovator state in the field of public policy, and legislative sponsors in other states have introduced bills that use California's definition of cloning. If these bills pass, California's definitional error will be repeated. Senate Bill 1602 in the U.S. Congress explicitly exempts egg cell nuclear transfer.

In addition to defining cloning in diverse ways and without extended discussion of cloning's variants, the states and Congress run the risk of passing laws that will be obsolete with surprising speed. If laws are worded loosely, they proscribe too much, but if worded narrowly, they encounter what LeRoy Walters has called the "instant obsolescence" of laws in the rapidly changing field of reproductive technologies.[23] To be sure, laws can always be amended; when lobbying for his cloning bill, Senator Frist said that if it inadvertently prohibited legitimate medical research, the law could be revisited.[24] Yet revisiting a law is awkward and not always successful, which leaves legitimate activities in a state of legal limbo.

A somewhat different approach to cloning would appeal to aspirational principles or voluntary limits that would guide action but leave "room for adjustment" as techniques change. In contrast to a law, which is a binding rule of conduct enacted by the government, aspirational documents present a framework for making decisions based on shared values and goals.[25] Such an approach appears in numerous settings worldwide, where bioethics commissions and government and nongovernment organizations show a shared cautionary sentiment about SCNT. For example, the World Medical Association, a group of seventy medical associations from around the world, has called for a moratorium, but not a ban, on cloning, as has the General Assembly of the World Health Organization, which resolved that cloning to replicate human individuals is "ethically unacceptable and contrary to human integrity and morality."[26]

An aspirational document of particular interest is the Universal Declaration on the Human Genome and Human Right, signed in 1997 by the 186 member nations of the United Nations Educational, Scientific, and Cultural Organization (UNESCO).[27] This document, nine drafts of which were written by a special bioethics committee of UNESCO over a four-year period, is designed to balance scientific advance with human dignity and rights. Called UNESCO's most significant document, it was prepared in time for the 1998 celebration of the fiftieth anniversary of the United Nations Declaration of Human Rights.

Drafting of the Universal Declaration started years before Dolly's birth, so the emergence of cloning posed unexpected controversies in 1997 when some nations advocated more direction in the document to list techniques to be prohibited, such as germline genetics and cloning. Eventually a compromise was reached and the declaration proscribed cloning in passing when it recommended that "practices . . . contrary to human dignity, such as cloning, shall not be permitted." This passage identifies cloning as the only technique specifically mentioned in the declaration, contrary to original intentions to build a principle-based rather than technique-based document. Moreover, the document does not define cloning or defend the conclusion that cloning is contrary to human dignity. Because the declaration is aspirational only, it is not legally binding on the 186 member nation-states that signed it.

A second international document comes from the Council of Europe, a group of forty states that opened a Convention on Human Rights and Biomedicine in 1997 for the signatures of its members.[28] Following news of Dolly's birth and Richard Seed's announcement, the council drafted a second protocol, the Additional Protocol to the Convention for the Protection of Human Rights and Dignity of the Human Being with Regard to the Application of Biology and Medicine, on the Prohibition of Cloning Human Beings, and opened it in early 1998 for the signatures of the member states that had signed the original Bioethics Convention.[29] It states that "any intervention to create a human being genetically identical to another human being, whether living or dead, is prohibited." The protocol defines cloning loosely: "For the purposes of this article, the term human being 'genetically identical' to another human being means a human being sharing with another the same nuclear gene set." Unlike the UNESCO Declaration, the Bioethics Convention and the Cloning Protocol will be legally binding on the nations that sign it.

Numerous other documents have been issued by government and nongovernment agencies such as the Parliamentary Assembly of the

Council of Europe and the Bioethics Committee of the Human Genome Organization. Some urge that cloning be banned; others rely on moral suasion and voluntary moratoria. All reveal serious concern about cloning's impact on individual dignity, but few have fleshed out this argument in detail.

At the level of individual nations, many governments are silent on cloning. Others have embryo research laws that would indirectly limit cloning and cloning research. For example, Norway forbids all embryo research, and its law is broad enough to prohibit cloning.[30] Still others have embryo research laws that explicitly bar cloning. Germany's Embryo Protection Act of 1990 makes it a criminal offense to engage in cloning by embryo or fetal cloning or SCNT methods.[31] Britain's embryo research and assisted conception law sets up a licensing authority that oversees research on embryos and gametes; as the law is interpreted, licenses may not be given for twinning or SCNT.[32] Other nations are weighing new laws to target cloning, including Switzerland, the Netherlands, France, Japan, Canada, Norway, and Italy. Outside Europe, most nations are silent on embryo research and cloning, although the state of Victoria, Australia, enacted a law proscribing cloning in 1984, which has been superseded by another law that also prohibits cloning.[33]

Despite concern about cloning, there is no single national approach to cloning; on the contrary, there are significant differences among nations created by differing values placed on the role of medical research in society, degree of trust in the scientific enterprise, prevailing ideas about the moral status of the embryo, values placed on the importance of assisted conception in society, and historical circumstances. Although the UNESCO and Council of Europe documents aim at achieving a "transnational harmonization,"[34] one can expect variation among nations as they define what is and what is not contrary to human dignity.

In summary, efforts to regulate cloning at the state, national, and international level revolve around whether cloning by SCNT should be banned. Many factors combine to suggest that this question is still premature, including the political context, the unexplored variations in cloning types, and the indirect way cloning will unfold in the context of human medicine.

Expanding the Policy Context

Given the gradual ways in which procedures central to cloning will emerge and public resistance will dissipate in light of the perceived medical benefits of the indirect advances, several observations can be drawn about lawmaking efforts related to cloning. First, cloning laws should not be crafted under a false sense of urgency. In writing about germline genetics in 1994, John Fletcher noted that "no other future possibility [than germline genetics] . . . has been the object of more bans, moratoria, [and] regulations" in the absence of a "real context."[35] The sudden news of SCNT in 1997 suggests the need to update this conclusion to read that no other prospect than cloning has been the object of as many bans, moratoria, and regulations in a short period of time and in the absence of compelling evidence that cloning is imminent. Although the cloning of mice and other species since 1997 moves the prospect of human cloning closer, scientific hurdles, political realities, and ethical reservations will put significant brakes on the process.

The felt need for quick action gears attention to single laws rather than multidimensional policies. Yet time exists to forge policies in which multiple interest groups play a central role, and temporary measures can be initiated to diminish the likelihood of cloning in the meantime. For example, three groups representing virtually all scientists and practitioners with the capacity to clone in the United States have called on their members to refrain from cloning.[36] In addition, the U.S. Food and Drug Administration has asserted its authority to ensure that human experimentation using SCNT to generate a child will not proceed until questions of safety are addressed.[37]

Second, there is merit in crafting a cloning policy rather than drafting a cloning law. To be sure, a national law against cloning has an appealing simplicity to it: One defines cloning, proscribes it, sets a penalty, and enforces it. Yet a ban on an isolated technique years before its imminence would set an unfortunate precedent for medicine and scientific research. Moreover, it would give a false sense of closure on something that is part of a much larger whole. For example, the excitement generated among scientists by the birth of a transgenic cloned lamb, calves, and other species suggests that cloning in the future may be used to facilitate germline genetic interventions rather than to generate more than one individual sharing the same genome. If cloning and germline interventions intersect, it makes sense to direct our attention to policies monitoring innovations in assisted conception in general rather than to technique-by-technique lawmaking.

A parallel can be drawn to policy relating to more standard assisted reproductive technologies such as embryo freezing and gamete donation. Although critics have called for laws each time an unusual technique arises, legislative action has been slow to emerge. In the meantime, professional associations, academics, and others have developed more comprehensive policy proposals that address issues common to many techniques, such as informed consent, parent-child relationships, and safety.[38] In addition, Congress passed a clinic reporting law, the application of which involved varied interest groups.[39] All this points out the distinction between punitive one-time laws and graduated policies and reveals how the latter can be crafted over time.

Third, an incremental policy can benefit from a clinical voice in which physicians, scientists, and potential patients contribute an empirical underpinning to aid public understanding about how cloning-related developments are likely to unfold. In addition, policy experimentation is useful in which rules for voluntary line-drawing are proposed and refined by specialists who share an interest in similar policy matters, including physicians, scientists, representatives of interest groups and professional societies, academics, policy analysts, and media experts.[40]

The logical community for overseeing cloning is one with an ongoing interest in assisted reproductive technologies, such as academics; members of professional societies such as the American Society for Reproductive Medicine and Federation of American Societies for Experimental Biology; nonprofit organizations such as NABER; advocacy groups such as RESOLVE, policymakers in such agencies as the Food and Drug Administration and Centers for Disease Control; and lawmakers who have introduced bills relating to assisted reproductive technologies.

The medical community can contribute a clinical perspective to the incremental emergence of steps that might lead to SCNT. For example, a rudimentary form of cloning would be twinning of embryos by blastomere separation. If a couple generated only one embryo, for example, the four cells of the embryo could be separated and nurtured to become individual embryos for transfer. Policies developed by the medical community might include the following matters of escalating difficulty in a clinic twinning policy:[41]

— Should human embryo twinning be permitted for nontherapeutic uses (e.g., to aid in preimplantation diagnosis or nontherapeutic research) in which the embryo twin is destroyed?
— Should human embryo twinning be permitted to increase the number of embryos for a couple who have documented troubles generating more than one embryo?
— If twinning is permitted to increase the number of embryos for couples who produce only one or two embryos, should embryo twins be frozen for later thawing and transfer in the event no pregnancy occurs?
— If twinning is permitted to increase the number of embryos for couples who produce only one or two embryos, should embryo twins be frozen for later thawing and transfer in the event a pregnancy and birth does occur, which would lead to spaced twins?
— If twinning is permitted to increase the number of embryos for couples who produce only one or two embryos, should frozen embryo twins be donated to another couple, at the progenitors' request, which might produce spaced identical twins living with different rearing parents?

Efforts to make such distinctions will help pinpoint what it is about genome replication that disturbs and where lines should and can be justifiably drawn. After embryo twinning it makes sense to examine embryo cloning by nuclear transfer whereby the nuclei of embryo cells are transferred to enucleated donor eggs. If the intention for twinning and embryo cell nuclear transfer is the same (generating more embryos for transfer), should the policy be the same, or does the difference in technique justify different rules for each?

Similar questions may be posed for fetal cell nuclear transfer, common in animal research but not yet debated for humans. Here fetal cells would be cloned to tap an abundant source of nuclei for infertile couples.[42] Cloning fetuses would not involve the eugenic premise of duplicating the genomes of known adults deemed genetically desirable, but it would raise disturbing issues.[43] In one scenario, a woman who terminated her pregnancy could donate the cells of the aborted fetus to infertile couples. In a second scenario, a couple might try to clone a spontaneously aborted fetus in an attempt to conceive and gestate a child with the same genome as the fetus lost to miscarriage. Assuming the miscarriage occurred for reasons other than a genetic or chromosomal abnormality, in which case a second pregnancy using the same genome would be ill-destined, the attempt to "replace" a lost fetus might be a technologically achievable request at some point in the future.

In a third scenario, a woman who intended to carry her pregnancy to term might request that fetal cells, which would be easier to clone than the differentiated cells of a child, be withdrawn for later cloning to generate a spaced identical sibling. She might consider this, for example, if she was unlikely to conceive with her own eggs again. Would the same arguments directed against somatic cell cloning apply also to fetal cloning? Might arguments apply to one of these scenarios more than others? Developing policies that accept or reject variations within cloning with reasons would help ensure defensible cloning policies.

A fourth observation about lawmaking and cloning is that principles developed in the international arena can provide guidance in the crafting of private and public sector policy. International documents reveal a powerful urge to protect human dignity and individuality, which the United States has acknowledged. Although not a member of UNESCO, for example, the United States played an active role in drafting the Declaration on the Human Genome and Human Right. Moreover, in 1997 a summit of leaders from eight major industrialized democracies met in Denver, and at that summit the leaders, including

President Clinton, pledged "close international cooperation" to prohibit cloning through SCNT.[44]

The perspective of the U.S. Congress on reproductive and genetic technologies is shaped by the nation's constitutional and political context and by its status as a technologically advanced nation with researchers, clinicians, and corporations that stand to gain from scientific research and development. Still, it would be arrogant to dismiss international sentiment. As in war, the environment, and human rights, no single country, including the United States, ought to act in splendid isolation. Debates over cloning policy will be richer if crossnational and international documents and principles are acknowledged and addressed.

Fifth, if lawmakers are intent on regulating cloning, they, as well as scientists, should be guided by a measure of humility. For scientists, humility means an awareness that what is done in the laboratory has societal repercussions. For lawmakers, humility means an awareness that prior restraints on science without compelling reasons would be unprecedented and of dubious value. Among other things, variations in cloning have not yet been explored. Critics claim, for example, that cloning will threaten individuality, but will that injury may be as compelling for embryo twinning as for adult SCNT? To examine the societal costs posed by varieties of cloning is to provoke the kind of precise thinking necessary for integrated policies. Until the rationale behind these techniques is recognized and integrated into policy schemes, preemptive laws will be punitive but not necessarily useful or flexible in light of changing science.

Conclusions

John Kingdon has likened public policy to a primeval soup in which "ideas are floated, bills introduced, speeches made; proposals are drafted, then amended in response to reaction and floated again."[45] Policy incrementalism is mirrored in technological incrementalism, in which serendipitous findings are discovered and seemingly unrelated discoveries take on dimensions of their own. Yet this dual incrementalism has been nearly obviated by the nature of the cloning debate. On the policy front, the orienting question is whether cloning ought to be banned. On the technology front, the orienting query is whether individuals ought to be able to replicate their genomes. In fact, however, "cloning" subsumes a variety of procedures and policy choices that include more than either banning the entire set of procedures or doing nothing.

This chapter outlined how the debate can be enriched before additional cloning bans are initiated. Given the indirect ways cloning will emerge, bans at the state or federal level will give a false sense of security to those opposed to cloning and will lessen the incentive to develop tightly argued restrictions on or approvals of cloning. In the event that cloning bans are later declared unconstitutional for their intrusion on scientific inquiry or reproductive liberty or are struck down for vagueness or overbreadth, an opportunity will have been lost to develop a set of working rules that effectively set limits. The presence of working rules, developed over time through empirical inquiry and with the help of the medical community, will arguably provide a more effective foundation for managing developments related to cloning than will targeted bans on a technique that has many incarnations left in it.

NOTES

This article draws from Andrea L. Bonnicksen, "Should Human Cloning Be Prohibited?: Integrating the Medical Community into Public Policymaking," presented at the 1998 Annual Meeting of the American Political Science Association, Boston.

1. According to the American Society for Reproductive Medicine, human cloning is "the practice of cloning an existing or previously existing human being by transferring the nucleus of an adult, differentiated cell into an oocyte in which the nucleus has been removed and implanting the resulting product for gestation and subsequent birth." Letter from J. Benjamin Younger, Executive Director, American Society for Reproductive Medicine, to Senator John H. Carrington, Apr. 15, 1997 (on file with the author).

2. D. Solter, "Lambing by Nuclear Transfer," *Nature* 380 (Mar. 7, 1996): 24–25, reporting on S. M. Willadsen, "Nuclear Transplantation in Sheep Embryos," *Nature* 320 (Mar. 6, 1986): 63–65; I. Wilmut et al., "Viable Offspring Derived from Fetal and Adult Mammalian Cells," *Nature* 385 (1997): 810–13; L. Meng and D. P. Wolf,

"Nuclear Transfer in the Rhesus Monkey," *Abstracts of the 51st Annual Meeting of the American Society for Reproductive Medicine*, Seattle, Wash., Oct. 7–11, 1995, program supplement, p. S236.

3. Wilmut et al. "Viable Offspring."

4. T. Wakayama, A. C. F. Perry, M. Zuccotti, K. R. Johnson, and R. Yanagimachi, "Full-Term Development of Mice from Enucleated Oocytes Injected with Cumulus Cell Nuclei," *Nature* 394 (July 23, 1998): 369–74.

5. Arlene Judith Klotzko, "Voices from Roslin: The Creators of Dolly Discuss Science, Ethics, and Social Responsibility," *Cambridge Quarterly of Healthcare Ethics* 7 (Spring 1998): 121–40; Gina Kolata, "Holstein Calves Cloned from Cells, Paper Says," *New York Times*, Apr. 23, 1998, p. A9.

6. A. L. Bonnicksen, "Transplanting Nuclei between Human Eggs: Implications for Germ-Line Genetics," *Politics and the Life Sciences* 17 (1998): 3–10.

7. Gina Kolata, "Scientists Face New Ethical Quandaries in Baby-Making," *New York Times*, Aug. 19, 1997, p. B11; J. Zhang, J. Grifo, A. Blaszczk, L. Meng, A. Adler, A. Chin, and L. Krey, "In Vitro Maturation (IVM) of Human Preovulatory Oocytes Reconstructed by Germinal Vesicle (GV) Transfer," *Abstracts of the American Society for Reproductive Medicine*, Cincinnati, Ohio, Oct. 18–22, 1997, program supplement, p. S1.

8. Carol M. Warner, "Genetic Engineering of Human Eggs and Embryos: Prelude to Cloning," *Politics and the Life Sciences* 17 (1998): 33–34.

9. D. S. Rubenstein, D. C. Thomasma, E. A. Schon, and M. J. Zinaman, "Germ-Line Therapy to Cure Mitochondrial Disease: Protocol and Ethics of In Vitro Ovum Nuclear Transplantation," *Cambridge Quarterly of Healthcare Ethics* 4 (1995): 316–39.

10. See, for example, *Summary Report: Engineering the Human Germline* (Los Angeles: UCLA Science, Technology, and Society Program, Center for the Study of the Evolution and the Origin of Life, June 1998).

11. Elizabeth Pennisi, "Transgenic Lambs from Cloning Lab," *Science* 277 (Aug. 1, 1997): 631; Klotzko, "Voices from Roslin," 131–32.

12. J. L. Hall, D. Engel, P. R. Gindoff, G. L. Mottla, and R. J. Stillman, "Experimental Cloning of Human Polyploid Embryos Using an Artificial Zona Pellucida," *Abstracts of the American Fertility Society*, Montreal, Oct. 11–14, 1993, program supplement, p. S1; R. Voelker, "A Clone by Any Other Name Is Still an Ethical Concern," *Journal of the American Medical Association* 271 (1994): 331–32.

13. Ruth Macklin, "Cloning without Prior Approval: A Response to Recent Disclosures of Noncompliance," *Kennedy Institute of Ethics Journal* 5 (1995): 57–60; "Embryo Cloners Jumped the Gun," *Science* 266 (Dec. 24, 1994): 1949.

14. National Advisory Board for Ethics in Reproduction (NABER), "Report on Human Cloning through Embryo Splitting: An Amber Light," *Kennedy Institute of Ethics Journal* 4 (1994): 251–82; Ethics Committee, American Society for Reproductive Medicine, "Embryo Splitting for Infertility Treatment," *Fertility and Sterility* 67, suppl. 1 (1997): 4S–5S.

15. National Bioethics Advisory Commission, *Cloning Human Beings: Report and Recommendations of the National Bioethics Advisory Commission* (Rockville, Md.: National Bioethics Advisory Commission, June 1997), 66–67.

16. Eliot Marshall, "Biomedical Groups Derail Fast-Track Anticloning Bill," *Science* 279 (Feb. 20, 1998): 1123–24.

17. Ibid.; "Regarding: Legislation to Ban Cloning of Human Beings," <http://thomas.loc.gov/cgi-bin/query>; Nicholas Wade, "Senate Plans to Weigh Ban on Cloning," *New York Times,* Feb. 10, 1998, p. A16.

18. Marshall, "Biomedical Groups Derail Fast-Track Anticloning Bill."

19. S.B. 1602, 105th Cong. (1998).

20. S.B. 1601, 105th Cong. (1998).

21. H.R. 922, 105th Cong. (1997). See House Report 105–239 (Part 1), "Human Cloning Research Prohibition Act." U.S. House of Representatives, 105th Cong., 1st Sess. Committee on Science, Aug. 1, 1997.

22. Cal. Statutes 1997, chap. 688 (S.B. 1344), sec. 5.

23. LeRoy Walters, "Ethics and New Reproductive Technologies: An International Review of Committee Statements," *Hastings Center Report* 17 (June 1987): 3S–9S.

24. Lizette Alvarez, "Senate, 54–42, Rejects Republican Bill to Ban Human Cloning," *New York Times,* Feb. 23, 1998, p. A18.

25. Bartha Marie Knoppers, "Cloning: An International Comparative Overview," Commissioned Papers, *Report and Recommendations of the National Bioethics Advisory Commission,* Vol. II (Rockville, Md.: National Bioethics Advisory Commission, June 1997), G1–G13.

26. "Medical Associations Urge Restraint on Cloning Research," Feb. 28, 1997 (Paris: Reuter); Declan Butler, "Calls for Human Cloning Ban 'Stem from Ignorance,'" *Nature* 387 (May 22, 1997): 324.

27. Declaration on the Human Genome and Human Rights, <http://www.unesco.org/ibc/uk/genome/project/index.htm>.

28. Convention for the Protection of Human Rights and the Dignity of the Human Being with Regard to the Application of Biology and Medicine: Convention on Human Rights and Biomedicine, European Treaties, ETS No. 164 (1997).

29. Additional Protocol to the Convention for the Protection of Human Rights and Dignity of the Human Being with Regard to the Application of Biology and Medicine, on the Prohibition of Cloning Human Beings, European Treaties, ETS No. 168 (1998).

30. L. B. Andrews and N. Elster, "Cross-Cultural Analysis of Policies Regarding Embryo Research," ms. commissioned by the National Institutes of Health, 1994.

31. "German Embryo Protection Act (October 24th, 1990): Gesetz zum Schutz von Embryonen," *Human Reproduction* 6 (1991): 605–6.

32. Human Fertilisation and Embryology Act, reprinted in *International Digest of Health Legislation* 42 (1991): 69–85. See also Human Fertilisation and Embryology Authority, *Code of Practice,* 4th ed. (London: Human Fertilisation and Embryology Authority, 1998), 55–56.

33. The 1984 law is described in Margaret Brumby and Pascal Kasimba, "When Is Cloning Lawful?" *Journal of In Vitro Fertilization and Embryo Transfer* 4 (1987): 198–204. The law has been superseded by Australia (Victoria), Infertility Treatment Act 1995, no. 63 of 1995, date of assent June 27, 1995 (151 pp.), Ausl. (Vic) 97.1, reprinted in *International Digest of Health Legislation* 48 (1997): 24–33.

34. Bartha Maria Knoppers and Ruth Chadwick, "The Human Genome Project: Under an International Ethical Microscope," *Science* 265 (Sept. 30, 1994): 2035–36.

35. John C. Fletcher, "Germ-Line Gene Therapy: The Costs of Premature Ultimates," *Politics and the Life Sciences* 13 (1994): 225–27.

36. These groups are the American Society for Reproductive Medicine, the Biotechnology Industry Organization, and the Federation of American Societies of Experimental Biology; see "Regarding: Legislation to Ban Cloning of Human Beings."

37. See, for example, letter from Sharon Smith Holston, Deputy Commissioner for External Affairs, Food and Drug Administration, to Senator Edward M. Kennedy, Feb. 10, 1998. Inserted in 144 Congressional Record S562, Feb. 10, 1998.

38. See, for example, ISLAT Working Group, "ART into Science: Regulation of Fertility Techniques," *Science* 281 (July 31, 1998): 651–52.

39. P.L. 102–493, Fertility Clinic Success Rate and Certification Act of 1992, 106 Stat. 3146.

40. On the idea of a policy community, see Giandomenico Majone, *Evidence, Argument, and Persuasion in the Policy Process* (New Haven, Conn.: Yale University Press, 1989), 161.

41. This is developed in more detail in Andrea L. Bonnicksen, "Procreation by Cloning: Crafting Anticipatory Guidelines," *Journal of Law, Medicine, and Ethics* 25 (1997): 273–82, at 274.

42. Some clinicians have proposed using fetal eggs or ovaries to produce eggs for fertilization. See A. Shushan and J. G. Schenker, "The Use of Oocytes Obtained from Aborted Fetuses in Egg Donation Programs," *Fertility and Sterility* 62 (1994): 449–51.

43. For an examination of issues related to fetal egg use, see Andrea L. Bonnicksen, "Fetal Motherhood: Toward a Compulsion to Generate Lives?" *Cambridge Quarterly of Healthcare Ethics* 6 (1997): 19–30.

44. "Communique Adopted by the Summit of Eight," *Bulletin EU* 7/8 (1997).

45. John W. Kingdon, *Agendas, Alternatives, and Public Policies* (Boston: Little, Brown, 1984).

8 Human Cloning: Public Policy When Cloning Is Safe and Effective

Given the still rudimentary knowledge of somatic cell cloning techniques, few persons interested in cloning as a responsible means of family formation would proceed with cloning unless they had a reasonable basis for thinking that it would lead to the birth of physically healthy children. Of course, it is always possible that rogue scientists or individuals would attempt to clone humans outside of established channels, but that risk is no greater with cloning than with any other experimental technique.

An important set of policy issues will arise if animal and laboratory research shows that cloning is safe and effective in humans. Should all cloning then be permitted? Should some types of cloning be prohibited? What regulations will minimize the harms that cloning could cause? What body or entities should enact such regulations? Considering these questions is instructive for coming to terms both with human cloning and with genetic selection generally.

1. Restrictions on Who May Clone and Rear

If human cloning is shown to be safe and effective, an important set of policy issues will concern whether it should be banned totally or whether only certain uses of it should be permitted.

A. TOTAL BAN The National Bioethics Advisory Commission (NBAC) relied on considerations of physical safety in calling for a complete ban on human cloning, although it also intimated that any human cloning was morally unacceptable even if physically safe.[1] Once it is shown that adult cell nuclear transfer cloning is safe and effective in humans, its utility for couples faced with infertility or other needs and the likelihood that it will cause substantial harm if practiced will have to be squarely faced.

Focusing only on married couples intent on gestating and rearing their own children, several valid uses consistent with prevailing reproductive and family-formation practices have been identified. Cloning of embryos might make having children possible at all, and embryonic or somatic nuclear cell transfer might help a couple to obtain tissue for an existing child. Cloning might also enable a couple to continue the genes of a dead or dying child or spouse.

Other important uses of cloning exist for couples with gametic infertility or couples who are carriers of genes for severe genetic disease. If both partners lack gametes, they might want to use the DNA of an unrelated third party rather than embryo donations from unknown infertile couples. If one of them lacks gametes, use of their own DNA might be preferable to sperm from an unknown stranger or eggs purchased from an egg donor.[2] Although there are some differences in method, the goals sought here are similar to those sought in current assisted reproductive practices and raise similar issues. The choice to accept the special challenges of cloning by nuclear transfer in forming a family would seem to fall well within the range of discretion ordinarily granted to couples.

The analysis of harm supports this conclusion. Safety concerns are not relevant because we are assuming safety and efficacy. Eugenics is a larger concern that arises with most genetic selection technologies. Objectification concerns ignore the complex ways in which children are ends in themselves while serving other purposes as well. Individuality, autonomy, and lineage are important concerns, but cloning is consistent with respecting and nurturing the individuality and autonomy of cloned children. Indeed, it is hard to believe that parents would not be interested in promoting their child's individuality and uniqueness even if they have reasons for wanting the child to have a particular genome.

Nor does the risk of confusing kinship relations justify a ban on all cloning. Kinship issues significantly arise only when an unrelated third party or one of the partners provides the DNA. The first is typical of embryo donation—an accepted practice. The second is more novel, but there are good reasons for thinking that responsible parents could deal with the special challenges which intergenerational genetic identity might pose. The most troublesome case—cloning and rearing one's own genetic parent—is likely both to be rare and to be so divorced from the usual reproductive context that it might not be an exercise of procreative liberty at all.

When carefully analyzed, the alleged harms of cloning tend to be either highly speculative or moralistic or subjective judgments about the meaning of family and how reproduction should occur. Such choices are ordinarily reserved to individuals, free of government coercion or definition of what provides reproductive meaning.[3] One need not accept human cloning as a morally acceptable way of family formation. But personal moral opposition alone is not an adequate basis for laws that prohibit others from using a technique that enables them to achieve legitimate goals of having and rearing biologically related children. Given the general presumption in favor of reproductive freedom, a ban on safe and effective human cloning in all circumstances is not justified.

One can make the same points in constitutional terms. The right to have children and rear them has been recognized as a fundamental right.[4] Despite the lack of textual specification, the idea is widely accepted that coital reproduction, at least between married couples, is protected against state restriction unless compelling reasons for interference are shown. It should follow that infertile couples also have rights to have offspring, for they have the same interests in having and rearing offspring as the coitally fertile. Thus laws that interfered with their ability to use noncoital techniques involving their own gametes, such as in vitro fertilization (IVF) or artificial insemination with husband sperm, should have to meet the compelling interest standard to be valid. If a right of infertile couples to use noncoital means of reproduction is recognized, then that right might be plausibly extended to the use of donor eggs, sperm, and embryos (without necessarily extending to full surrogacy).[5]

A necessary implication is that some degree of genetic selection is also protected because the ability to select will often determine whether or not reproduction occurs. Thus the right to know one's carrier status and act on it either before or after conception, implantation, or pregnancy would follow. If some right to negative selection is presumptively protected, then some forms of positive selection should be as well, for that too will often determine whether reproduction occurs.[6] Thus embryo splitting and nuclear transfer cloning fall into the presumptively protected category, for their availability may well determine whether or not a couple has and rears biologically related children. If so, restrictions on cloning would be valid only if necessary to prevent tangible harm to others. As we have seen, however, most of the objections to cloning do not meet that standard. They are either too speculative or too moralistic to justify interference with quintessentially private choices about family.

Although a cogent argument for a constitutional right to form biologically related families, including the use of cloning when necessary, can be articulated, it may be naive to expect all courts to accept such reasoning at the present time. The right to procreate as an aspect of procreative liberty has rarely been litigated as such. Important distinctions about types and ways of fulfilling procreative desires have not entered public discourse. The Supreme Court is increasingly reluctant to recognize new fundamental rights as part of Fourteenth Amendment substantive due process.[7] However, even if courts do not recognize a constitutional right to clone, policymakers and professionals should nevertheless acknowledge that the interests and values underlying other reproductive decisions provide a strong basis for a right to use assisted reproductive and genetic selection techniques, including cloning, to complete the family project, and respond accordingly.

But this argument only shows that a ban on all cloning, including the family-centered uses described above, is overbroad. We must also ask whether there are some uses of cloning that should not be permitted, and whether some regulation of permitted uses is desirable once human cloning is medically feasible.

B. NO CLONING WITHOUT REARING   A ban on human cloning unless the parties requesting the cloning will also rear is a much better policy than a ban on all cloning. The requirement of having to rear the clone addresses the worse abuses of cloning. It prevents a person from creating clones to be used as subjects or workers without regard for their own interests. For example, situations like those in *Boys from Brazil* or *Brave New World* would be prohibited because the initiator is not rearing. This rule will assure the child a two-parent rearing situation—a prime determinant of a child's welfare.[8] Nor would it violate the initiator's procreative liberty because merely producing children for others to rear is not an exercise of that liberty.[9]

Ensuring that the initiating couple rears the child given the DNA of another prevents some risks to the child, but still leaves open the threats to individuality, autonomy, and kinship that many persons think that cloning presents. I have argued that parents who intend to have and rear a healthy child might not be as prey to those concerns as feared, yet there is still a chance that some cloning situations, because of the novelty of choosing a genome, might produce social or psychological problems.

Those risks should be addressed in terms of the situations most likely to generate them and the regulations, short of prohibition, that might minimize their occurrence. It hardly follows that all cloning should be banned because some undesirable cloning situations might occur. Like other slippery slope arguments, there is no showing that the bad uses are so likely to occur, or that if they did, their bad effects would so clearly outweigh the good, that one is justified in imposing the loss of the good in order to prevent the bad.

C. BAN ON REARING A CLONE OF SELF   An important policy question that the safety of human cloning will present is whether self-cloning—using the DNA of one of the rearing partners to create the child whom they rear—should be permitted. The concern is the confusion in kinship and rearing roles that cloning and rearing a clone of oneself might produce (as well as the confusion it causes for the clone source's genetic parents). In this view, cloning of one's own children or an unrelated third party is acceptable, at least when the initiator also rears, but cloning oneself is not.

The wisdom of this policy depends on an assessment of the need or benefit cloning and rearing one's own clone serves and the harm it poses. The reasons for self-cloning and rearing have been discussed. It is a plausible response to the need for gamete or embryo donation due to infertility or genetic disease, and depending on how one views the importance of genes in defining reproduction, is also highly reproductive. Of course, such a situation is likely to be highly fraught psychologically, but there is no reason to think that it will invariably lead to bad outcomes. As long as the rearing couple is aware of the dangers and is committed to giving the child a separate identity, the risks of harm may be minimal. Certainly they would be within the range of comparable risks assumed by parents in having children, or in exercising or failing to exercise genetic selection in other circumstances.[10]

Given the close connection with actual reproduction, one could also plausibly argue that a constitutional right to engage in genetic selection, as an aspect of the right to reproduce, should include the right to create and rear a child with one's own nuclear DNA. The mere prospect of harm or the novelty of the situation is not a sufficient basis for interfering with personal decisions about how one acquires genetically related children for rearing. In the strongest case, the couple will have no alternative way to rear a child genetically related to the spouse with gamete problems. If a couple is well-informed of these risks, is stable, and is interested in the well-being of the resulting child, their choice to rear a clone of one of the partners will be hard to distinguish from protected reproductive experiences, and should be treated accordingly.

D. BAN ON CLONING AND REARING ONE'S PARENT   The situation of gestating and rearing of a child formed from nuclear transfer from one's parent presents a less compelling case. Such cases are likely to be rare, but they may occasionally arise. This situation presents the maximum threat to traditional notions of kinship, for it completely reverses the intergenerational meaning of parent and child. The result-ing child may not be intrinsically harmed, for it has no other way to be born. It is also possible that the rearing parents will adapt smooth-ly to the psychological novelty of the situation and minimize their iden-tification of the child with the parent.

Still, the situation is so different from other cloning situations that the couple seeking a parental source of DNA should have the burden of establishing its bona fides as a method of family formation. It may be that the couple will be able to carry this burden. After all, daughter-to-mother oocyte donation and father-to-son sperm donation have occurred without apparent damage.[11] A ban on such cloning would not greatly interfere with the use of cloning in other situations. Unless the couple could show how parental cloning relates to a valid reproduc-tive project, it might not be fit within prevailing understandings of procreative liberty and therefore need not be protected.

E. BAN ON SINGLE WOMAN OR MAN CLONING   Some persons have suggested that cloning should be permitted only if it occurs in a two-partnered married setting.[12] Such a position would affect gay and lesbian couples and single men and women, who might choose to clone and rear themselves or another. There is a strong reproductive interest in at least some of those situations, which such a restriction would infringe. If that interest is to be respected, the policy should be modified to allow cloning in unmarried situations involving a two-partner com-mitted relationship, as might occur with a lesbian couple considering having a child by donor sperm or adoption.

This modification would still ban a single woman from rearing a clone of another or herself, or a single man from commissioning the clon-ing of himself or another, whom he will then rear alone. One could ar-gue that such a ban interferes with the reproductive or family formation rights of unmarried persons. The resolution of this question should de-pend upon whether single persons have rights or access to other forms of assisted reproduction and genetic selection. If single persons are grant-ed access to those other techniques, the question then posed is whether cloning presents such additional rearing problems that it should be treat-ed differently.

2. Regulation

If human cloning proves to be safe and effective and is permitted for some or all of the indications discussed, it is essential that it be done in ways that minimize the special risks that it poses. Some form of regulation seems desirable, if only the specification of guidelines for how it should optimally occur. As long as such regulations do not unduly burden access to the technology, they would not interfere with the right of couples to use cloning to form families.

A. CONSENT OF THE CLONE SOURCE   A key regulatory issue is whether the clone source—the source of the DNA—must consent to the cloning. This issue does not arise when embryos are cloned, though it is relevant when an existing child is cloned (or a previously cloned embryo is placed in the uterus and a later identical twin to an existing child is born).[13] However, it does not follow that the child's consent to use of his or her DNA is necessary. Children do not ordinarily have the right to determine whether their parents have additional offspring. The fact that the new offspring will be an identical twin does not create such a right, for that fact alone does not create additional burdens or problems for the existing child. As long as the DNA is obtained noninvasively from the first child, the parents should be free to use the DNA without that child's consent when the child is a minor.[14]

Somatic cell nuclear transfer with DNA from an existing person other than a minor child, on the other hand, should require the consent of the person whose DNA is used if he or she is alive. This is clearest if the DNA is obtained from him or her directly, for an unconsented-to touching or battery might be involved. It is possible, however, that techniques for recovering DNA from a person would involve such minimal touching that tort doctrines of battery would not apply.[15]

The consent of the clone source should also be required where the DNA is shed involuntarily in the course of living and recovered from benches, chairs, doorknobs, utensils, clothes, saliva from postage stamps, and other objects that a person has touched.[16] Although a person has no physical connection with his or her DNA once it is shed and implicitly abandons it by moving through the world, there are good reasons for assigning the person a limited property right to control whether his or her DNA is used to create another person.[17] Even if they have no direct legal rights or duties toward their later-born twin, the relationship is a novel one with potential psychological complications that persons should be free not to incur. A legal remedy in damages should be available for those whose DNA is used without their consent.[18] However, they should not thereby automatically qualify for rearing rights and duties in the resulting child.[19]

The requirement of consent of the clone source raises two further questions. One is the question of whether one may sell the right to use one's DNA.[20] Celebrities may try to reap additional income from the sale of their DNA to ardent fans or others. Because the demand for their DNA is a result of investment in their own human capital, one could argue that compensation is appropriate as an incentive to develop that capital fully.[21] Payment could also be justified as advance compensation for any social or psychological complications that arise from having a later-born identical twin. However, there is likely to be wide repugnance to the idea of selling DNA for cloning, just as there is repugnance to selling embryos or organs for transplant.[22] If adult somatic cell nuclear transfer becomes an accepted technique, one might expect laws to be passed that protect the clone source's right to consent to cloning but which prohibit payment or a market in DNA.

Another question concerns use of DNA of persons who are deceased, if such DNA proves viable for cloning.[23] Ordinarily, privacy rights expire with one's life, though certain interests in name and personality survive death.[24] It may be that DNA, like the rest of the body and its parts, will be subject to the wishes of the person while alive or next of kin upon death. Under this approach, the next of kin would have to consent to the use of the deceased's DNA as a source of nuclear transfer. This may not be unreasonable, given that a child born of the deceased's DNA will be an identical twin of the deceased and will have kinship relations with his or her survivors.

B. CONSENT OF THE CLONE SOURCE'S GENETIC PARENTS   Should the consent of the clone source's genetic parents also be required for cloning to occur? The argument that it should is based on its repro-ductive implications for them. Cloning creates a later-born identical twin for the clone source, but an additional genetic offspring for the clone source's genetic parents. Even if no rearing rights or duties in the resulting child attach, the social and cultural meanings that at-tend gene transmission may still stir up strong psychological and social feelings in the clone source's parents.

Because it is so like reproduction, a cautious societal policy toward cloning would give the clone source's genetic parents, as well as the clone source, the right to veto the use of the clone source's DNA for reproductive cloning. Such a policy could, however, lead to conflicts between the clone source and his or her parents over whether the DNA may be used to create another person. Ordinarily parents do not have a right to determine whether their offspring reproduce, and thus make them grandparents. Genetically, however, cloning of their offspring makes them parents, even though they may occupy the social role of grandparent.

If human cloning becomes an accepted technique, such conflicts will need resolution. At bottom the question is whether reproduction *tout court* through cloning is such an imposition on persons that their consent should be required, even if it will prevent their own offspring from transmitting their DNA to others. The answer will ultimately depend upon our assessment of the psychological burdens or mean-ing that attach to being cloned and to reproduction *tout court*.[25] One could reasonably view the interest of the clone source's parents as too intangible or tenuous to justify overriding their adult offspring's wishes to donate his or her DNA to others, just as the consent of both identi-cal twins is not necessary for one of them to reproduce, even though the resulting child will be the genetic child of the nonconsenting twin as well.[26] But then no rearing duties, such as child support or even con-tact, should be imposed on them as a result of the genetic connection of parenthood that will result.

C. INFORMED CONSENT AND PSYCHOLOGICAL SCREENING Hu-man cloning, though continuous with other assisted reproductive and genetic selection techniques, does pose special medical, psychologic, and social challenges that persons considering it should clearly understand.[27] Regulations to ensure that couples considering cloning are fully aware of the medical risks of the DNA transfer procedure itself, including likely success rates, are clearly desirable and justified. In addition, physicians or others providing DNA transfer procedures should inform couples contemplating such procedures of the special social and psychologic challenges posed so that the problems of individuality, autonomy, and lineage that could arise are minimized.

Wise policy would also require couples requesting cloning, particularly of themselves or their children, to undergo psychological screening and counseling before the procedure. This will ensure that couples are fully informed of those risks and will enable doctors to screen out those who seem unstable or not able to handle the special challenges posed. Such counseling should address the basis of the parents' desire to choose particular DNA for their expected child and the issues and problems that choice poses, including the importance of respecting the child's individuality and autonomy in his or her own right, and the dangers of parental expectations of the child based on the chosen genome.

D. REARING RIGHTS AND DUTIES IN RESULTING CHILDREN
An important regulatory issue that will arise with human cloning is the need to clarify parental rights and duties in resulting offspring. This will minimize detrimental legal battles over child custody and visitation and help reduce any confusion over kinship. Clarification and certainty can be achieved either by legislative specification of those relations or by legislative or judicial recognition of the precloning agreements of the affected parties.

The cloning of embryos or children does not pose this problem because the clone initiator and rearing parents are also the genetic parents. The relation of parent and child, grandparent and child, and child and sibling is clear because the child has the same kinship relations with those parties as if he or she had been coitally conceived. Still, the novelty of being the identical twin of an older sibling may affect sibling relations and to some extent how family members, if they are aware of a child's cloned origins, perceive them. Legislation specifying rearing relationships is least needed in this situation.[28]

Clarification of kinship relations is most needed when DNA from an unrelated third party or from one of the rearing partners or a member of his or her family is used. In the case of an unrelated third party, the standard case will pose the same kinship issues that arise with human embryo donation. The intention is for the recipients of embryo donations to gestate and rear a child who is genetically the offspring of others. Legislation in a few states has recognized this arrangement, and courts in other states are likely to reach the same result if ever presented with a dispute.[29]

A similar solution should apply to a child born after DNA transfer from a third party. One who provides the DNA for a child to be reared by another should be viewed as voluntarily relinquishing any rearing rights or duties in resulting children, as would the donor of the denucleated eggs into which the cloned DNA is placed. Law and regulation should give effect to this arrangement so that ensuing kinship relations are clear and certain. The law should also bar the clone source's genetic parents from any rearing role, even though the resulting child is their genetic offspring.

Kinship and rearing relations in the likely situation of individuals using their own DNA in lieu of gamete donation should also be specified. The person consenting to be cloned is undertaking to rear the resulting child. This commitment to function as legal father or mother, together with gestation, should be given binding legal effect just as it is with donor gametes.[30] The social situation makes the clone source the parent of the clone, who is the social parent's own later-born twin. The source's parents, who are the clone's genetic parents, should also be clearly recognized as social grandparents and not be assigned any rearing rights or duties.

Of course, the child, if informed of his or her cloned origins, will still have to deal with the fact that he or she has an identical twin in the world. Clarifying the legal status of the participants will help normalize the relationship. The importance of the genetic choice that made the child should recede as the child's individuality emerges. Questions of whether the child should be told of his or her origins and allowed to meet the source—the earlier-born twin—and that twin's genetic parents will also have to be addressed.[31] Overall, however, specification of resulting kinship relations and the consequent child-rearing rights and duties will minimize the social and relational problems that human cloning presents.

E. LIMIT ON NUMBER OF CLONES   Any practice of human cloning should have clear legal limits on the number of children who may be born with the same DNA, whether at the same or different times. Such a policy will minimize the problems of individuality that cloning appears to present for resulting children and will help ensure that any resulting child is fully regarded as a worthy individual in his or her own right.

A limit of no more than three clones born from DNA is a reasonable line, though other numbers might also reasonably be chosen. A three-child limit is consistent with most reproductive needs, and would even permit a couple to have simultaneously born identical twins by cloning. At the same time it would prevent the birth of many individuals with the same DNA, thus diminishing the popular perception of cloning as a way to produce multiple copies of a single person. Problems of ensuring individuality and separateness will remain, but they would be lessened if the couple and child knows that there is only one or two others with the same DNA.

Of course, the fourth, fifth or nth clone of particular DNA would not itself be harmed, not having any alternative way to be born. Nor would it hurt the children previously born with that DNA, for they gain no rights or duties as a result and may never even learn of later cloning.[32] The interest in having and rearing children is not involved in simply providing DNA for others to use in having offspring. On the other hand, couples seeking DNA that has already produced three additional children can argue that they will not gestate and rear children unless they can use the particular DNA that has already produced three children. Whether this claim deserves respect will depend on the importance assigned to having yet another child with the same DNA when several persons with that DNA already exist.

Professional guidelines for gamete donation now limit the number of children that can be born from one sperm donor to ten.[33] The same number has been suggested as a limit for children born from a single egg donor. The reason for these limits is to protect the egg donor and to minimize the risks of consanguineous marriages occurring unwittingly between people who are half-siblings. With cloning, the rationale for a limit is to minimize the risks to the individuality of the resulting children. A limit on the number—whether two, three, or *n*—will help, even though it will not eliminate all issues of separateness and uniqueness that using the DNA of others to have children poses.

F. PROFESSIONAL DISCRETION OR A REGULATORY AGENCY   An important issue in any regulatory approach is whether the professionals directly involved can be trusted to provide cloning or other reproductive services in a safe and ethical way, or whether a government agency is needed to oversee their practice. Some countries, such as Great Britain, have chosen the latter course and created a Human Fertilisation and Embryology Authority to regulate both the safety and efficacy of assisted reproduction and the introduction of new techniques.[34]

The United States, on the other hand, has a much more decentralized system that leaves considerable discretion to the professionals and consumers directly involved and the institutions where services are provided. Some federal regulation exists through the institutional review board system of reviewing research with human subjects, and FDA approval of the safety and efficacy of drugs, devices, and biologics. Voluntary moratoria have, from time to time, been respected by geneticists, and national commissions, such as the NBAC, have provided guidance to policymakers.[35] However, many persons think that a permanent federal commission or agency to oversee new reproductive and genetic technologies, such as cloning, is necessary. A major issue for future policymakers is whether the current decentralized system, where much discretion rests with the professionals and consumers directly involved, will continue when cloning and other genetic alteration techniques become available. Based on past experience, considerable reliance on professional discretion is likely to remain, though it is possible that the special concerns that arise with cloning could lead to a more permanent government body overseeing new reproductive and genetic techniques.

G. OTHER REGULATORY ISSUES   If human cloning becomes available to couples who need it, other regulatory issues will arise. A pressing one is whether a national registry of children created through DNA transfer should be created so that the incidence and effects of nuclear transfer cloning can be studied. Still another would be limits on the use of third-party DNA, for example, because of the possible effects on children or others from having particular DNA. As the technology of extrauterine gestation develops, the possibility of total gestation outside a uterus will also have to be addressed. Funding and access issues will also be important. An additional set of regulations to ensure health and safety will arise if cloning becomes a viable technique for producing tissue and organs for transplant from embryonic stem cells or early abortuses.

The Lessons of Cloning for Genetic Alteration

Our investigation has focused on human cloning as a potential technique for forming families. Issues of genetic selection and family autonomy, however, arise in many other contexts. Genetic selection through preconception carrier screening, preimplantation genetic diagnosis, or prenatal screening of fetuses now occurs in most pregnancies.[36] Germline gene therapy and other forms of positive intervention, such as nonmedical enhancement and intentional diminishment, loom on the horizon. Although they do not replicate the entire genome, they will make specific alterations of genes possible.

The current focus of such efforts is on gene therapy: inserting or deleting genes in order to cure or treat disease. A major obstacle has been developing a vector that can target genetic changes without causing other effects.[37] Once such vectors are developed for somatic cells, attempts to cure the disease at the embryo level will occur. If successful, the genetic alteration will pass on to that person's progeny.

The development of germline gene therapy will open the door to other attempts to insert, delete, or alter genes in early embryos or in cells that are then used as the source of DNA to create embryos.[38] Genetic alteration might be feasible if single genes are associated with particular characteristics, a highly unlikely but not impossible occurrence. If so, some parents will want to insert genes that enhance or increase intelligence, memory, beauty, strength, or other desirable traits in offspring. In rare cases, parents may even request prebirth genetic elimination of characteristics, such as deaf parents who wish the child they rear also to be deaf.[39] If these techniques become available, the scope of parental rights to select genetic characteristics of offspring will also arise.

The ultimate handling of issues of prebirth genetic alteration will depend on the precise alteration at issue, its benefits and risks, and how it fits into family and reproductive life generally. The discussion of cloning foreshadows many of those issues and shows where the ethical fault lines are. As with cloning, distinguishing good and valid uses from abusive or harmful ones will be key. A crucial issue will be whether the alteration is being used in a family-centered positive way or is being used to objectify or customize children without concern for the interests of the resulting child.[40] In resolving these issues, we will be defining the values at stake.

The question of altering genes to enhance (or diminish) offspring traits does differ from cloning in several important ways. First, most instances of genetic alteration will probably involve a couple that is reproducing genetically, if not also gestationally and socially, thus eliminating the need for another's consent, as is necessary in cloning a third party.[41] Second, issues of uniqueness and individuality will not be as salient because no replication of the DNA of another will be occurring.[42] Third, concerns about kinship are also likely to be absent here, for alteration can and is most likely to occur with an embryo or child formed from the couple's gametes (unless the DNA of another is being altered). Fourth, couples will usually choose to clone because of difficulties in having children in other ways.[43] Those using alteration may be able to reproduce coitally, but have chosen to go through IVF in order to alter genes.

Common to both cloning and gene alteration, however, is the issue of instrumentalization and objectification of the child. A pervasive concern about cloning is the risk that choosing the child's DNA will turn the child into an instrument or object to satisfy parental agendas that are unrelated to the child as an end in himself or herself. Although responsible families can successfully negotiate those dangers, the danger is a recurring one and is likely to exist in situations of genetic enhancement and diminishment as well.

A couple that seeks to alter the genes of an embryo to enhance an otherwise normal child's capabilities is in danger of being more interested in genes than in the child for his or her own sake.[44] Instead of conceiving coitally, they will undergo one or more expensive cycles of ovarian stimulation, egg retrieval, and IVF in order to gain access to embryos and their genes. It may be that parents would be motivated to take such steps out of love and concern for their expected child. On the other hand, it is also possible that few parents would go to such trouble unless they were fixated on the child as an object to serve their own needs. Given these possibilities, couples seeking to alter genes will have to convince gatekeepers and policymakers why the alteration is justified as part of their liberty interest in having or rearing offspring. Justifying intentional diminishment will be even more difficult.

Resolving these questions will pose many of the same constitutive questions that arise in determining the response to human cloning. Decisions about prohibiting all or certain types of genetic alteration will have to be made, and the effects of maldistribution of the technology considered. The extent of procreative liberty will also have to be defined, as will the notion of harm when children would not otherwise have been born but for use of the technique in question. Where cloning forces us to confront the meaning of identity and family, gene alteration will force us to confront the limits of instrumentality in seeking the good of offspring. As with cloning, the answers to these questions will constitute or define the very values and rights at issue.

Conclusion

Cloning and reproductive technology force us to think deeply about the meaning of genes, identity, reproduction, parenting, children, and our connection with family and nature. Such issues come to us structured as problems of liberal decision making. Is fundamental reproductive liberty involved? Do the harms justify intruding on those liberties? Is regulation to minimize ill effects consistent with the liberty rights at stake?

As we have seen with human cloning, and will see again as other techniques of genetic selection and manipulation become available, the answers to these questions are only partially determined by our past practices and understandings. Past practices will help form a bridge to the twenty-first century's genetic practices, but we will construct most of the bridge as we proceed. Cloning and genetic alteration will force us to define and constitute ourselves as we confront the genetic meanings of family and reproduction.

NOTES

This chapter previously appeared as pp. 1435–36 and 1439–56 of "Liberty, Identity, and Human Cloning," 76 *Texas Law Review* 1371 (1998), © 1998 by the Texas Law Review Association, and is reprinted with the permission of the *Texas Law Review* and John A. Robertson.

1. National Bioethics Advisory Commission, *Cloning Human Beings: Report and Recommendations of the National Bioethics Advisory Commission,* 1997, pp. 79–81.

2. An egg donor, however, might still be needed to provide the denucleated egg into which the cloned DNA would be placed.

3. One is reminded of the statement in *Casey v. Planned Parenthood,* 505 U.S. 851 (1992): "At the heart of liberty is the right to define one's own concept of existence, of meaning, of the universe, and of the mystery of human life. Beliefs about these matters could not define the attributes of personhood were they formed under the compulsion of the state."

4. See *Skinner v. Oklahoma,* 316 U.S. 535, 541 (1942). As Justice Brennan noted in *Eisenstadt v. Baird,* 405 U.S. 438, 453 (1972), "If the right of privacy means anything, it means the right to decide whether or not to bear or beget a child."

5. See John A. Robertson, *Children of Choice: Freedom and the New Reproductive Technologies* (Princeton, N.J.: Princeton University Press, 1994), 35–40.

6. See John A. Robertson, "Genetic Selection of Offspring Characteristics," *Boston University Law Review* 76 (1996): 424–29, quote on 421.

7. See *Bowers v. Hardwick,* 478 U.S. 186 (1986); "*Washington v. Glucksberg,*" 117 S.Ct. 2258 (1997).

8. James Q. Wilson, "The Paradox of Cloning," *Weekly Standard,* May 26, 1997, pp. 23–27.

9. First, if he is arranging for the clone of a third, as in *The Boys from Brazil,* and is not cloning himself, he is not reproducing. Second, the interest in helping others rear or choose the genome of whom they rear is not a fundamental interest or right. The situation is different if the initiator is acting as the agent of a couple who will rear. But then it is the couple's interest that is at stake, not that of the initiator as such.

10. For example, persons are now free before conception not to undergo screening to determine whether they are carriers of severe genetic disease, nor are women required to have fetuses screened for genetic disease, even though those actions could prevent the birth of a disabled or diseased child.

11. Lorna A. Marshall, "Intergenerational Gamete Donation: Ethical and Societal Implications," *American Journal of Obstetrics and Gynecology,* in press.

12. Wilson, "Paradox of Cloning."

13. It could be an issue when the cloned embryo is transferred to the uterus because that could create a later-born identical twin for the existing child.

14. This situation may be contrasted with parents who wish to test a child for a late-onset genetic diseases, such as Huntington disease, or the BRCA1 gene for breast cancer. Ethical standards now require that no testing of minors occur without their consent because of the heavy impact that such knowledge could have on them later in life. Having a later-born identical twin does not pose the same risks of harm to the child.

15. Even a slight brush might be sufficient to dislodge DNA from skin or hair from which somatic cell nuclei could be recovered.

16. Malcolm Ritter, "People Trail DNA Behind Them, Researchers Say," *Austin American Statesman,* June 19, 1997, p. A17.

17. Fourth Amendment law, however, allows the government to search garbage and trash cans on the street before pickup on the theory that there is no reasonable expectation of privacy in one's trash. *California v. Greenwood,* 486 U.S. 35 (1988). This precedent would allow the state to recover DNA from these sources for law enforcement purposes.

18. The leading case on rights in one's cells, *Moore v. Regents of the University of California,* 51 Cal.3d 120, 271 Cal.Rptr. 146,793 P.2d 479 (1990), denies a property right in spleen cells but recognizes a right of informed consent to the taking of the cells when the doctor has a fiduciary interest in their use. That case could accommodate damages for unconsented cloning *tout court* on a property or informed consent theory.

19. Note that the argument is not a constitutional one. At most reproduction *tout*

court is occurring, and the interests implicated may not warrant independent constitutional protection.

20. I am indebted to an unpublished paper by Neil Netanel on the right of publicity for elucidation of this issue.

21. However, the purchaser of the celebrity's DNA for cloning should remember that the source's investment in developing talent may be more responsible than DNA for his or her success.

22. See, for example, National Organ Transplant Act, sec. 301, 42 U.S.C. sec. 274e (1994), making it a federal felony to buy and sell human organs.

23. Unless the person is only recently deceased, his or her DNA may no longer be living and thus not be able to serve as the source of nuclear transfer.

24. Kenneth E. Spahn, "The Right of Publicity: A Matter of Privacy, Property, or Public Domain?" *Nova Law Review* 19 (1995): 1013, 1036–38, observes that whether a person's rights in his likeness or identity are descendible depends on whether the court interprets such interests as privacy rights, which terminate at death, or property rights, which pass to the estate.

25. A similar issue arises in disputes between divorcing couples over disposition of frozen embryos. See *Davis v. Davis,* 842 S.W.2d 588 (Tenn. 1992).

26. The nonconsenting twin will have a genetic heir of the same degree as if he or she had chosen to reproduce. I am indebted to Einer Kluge for this example.

27. A right of confidentiality should also apply. Information that someone has been cloned or is the product of cloning is important private information that should be zealously protected in medical or other records.

28. If comprehensive cloning legislation is being passed, then clarifying this point might be worthwhile. In addition, if the egg source is a donor, legislation that bars her from any rearing rights or duties would also be desirable.

29. Tex. Family Code #151.103. See also John A. Robertson, "Ethical and Legal Issues in Human Embryo Donation," *Fertility and Sterility* 64 (1995): 885–94.

30. In sperm donation, the consenting husband is the legal father for all purposes. A similar result should be recognized with a person whose DNA is used to produce a child whom that person will rear.

31. See Robertson, *Children of Choice,* 123–24.

32. One assumes that the fact of somatic cell nuclear transfer as the source of a child's DNA will be kept confidential. Even if the child is eventually told, there is no reason why later uses of the same DNA will be revealed to children or their parents. Of course, they may meet other children born with the same DNA in the normal course of life, and thus might learn in that way.

33. American Fertility Society, "Guidelines for Gamete Donation," *Fertility and Sterility* 59 (1993): 1S–9S. It is unknown how this guideline is enforced. Presumably it would be admissible as proof of the standard of care in a malpractice action against a clinic that so overused a particular donor that offspring inadvertently married their half-siblings.

34. Human Fertilisation and Embryology Act, 1990, chap. 37. Under the autho-

rizing legislation, cloning of an "embryo" is not permitted. See Pat Walsh and Andrew Grubb, "I Want to Be Alone," *Dispatches,* Spring 1997, 1–6.

35. NBAC Cloning Report, 96.

36. It is estimated that more than 60 percent of pregnancies in the United States are now screened for some kind of genetic malformation.

37. W. French Anderson, "Prospects for Human Gene Therapy," *Science,* Oct. 26, 1984, pp. 401, 407 (discussion of vectors for gene therapy and problems of genes rearranging their own structure or exchanging sequences with other retroviruses).

38. The report by the Wilmut team of the successful creation of a sheep with a human gene from insertion of genes into fetal skin cells indicates that similar genetic alteration before transfer of nuclear DNA will be feasible in humans. See Gina Kolata, "Lab Yields Lamb with Human Gene," *New York Times,* July 25, 1997, p. A18.

39. Selection of a genome with defective genes or genes that are not associated with prior traits would be a variation on intentional diminishment. Although strictly speaking not a case of diminishing what would otherwise have been a normal birth, this would be an example of not choosing a reasonable or good or best possible genome for someone. It would implicate a possible duty to make the child as well off as reasonably possible.

40. Although even customized children may not be harmed in being born if there were no other way for them to have come into being, the willingness of a couple to do so will be relevant to an assessment of whether they are validly exercising procreative liberty.

41. Unless they are seeking to alter the DNA of another in the process of nuclear transfer cloning. This might occur even if they have chosen a third party's DNA because of its attractiveness and might also occur if one's own DNA has been chosen.

42. However, autonomy and expectation issues would still arise because altering genes assumes that they are influential, if not also determinative. Rigid parental expectations could also arise.

43. Cloning to produce tissue is an exception.

44. Sought-after traits might include height, memory, intelligence, skin color, or certain behavioral traits. See Leroy Walters and Julie Gage Palmer, *The Ethics of Human Gene Therapy* (New York: Oxford University Press, 1997), 108–24.

9   Cloning and Public Policy:
Biotechnology Lessons

The relatively brief history of modern biotechnology has been marked with a series of public policy debates not unlike the current debate over human cloning. I write from experience as director of a program that promotes biotechnology research and is also charged with assessing the impact of biotechnology developments in the public arena, which most often has meant the public policy arena.

California is the birthplace of modern biotechnology. Recombinant DNA (gene splicing) techniques were developed here by scientists at the University of California at San Francisco and Stanford University. California is also the birthplace of commercial biotechnology, with a third of all U.S. companies found in the state.

California has also been the reluctant host to heated debate over biotechnology. It started in the early 1970s with the historic Asilomar conference, which led to a self-imposed moratorium on gene splicing by the scientific community. It continued in the 1980s with protests over the first outdoor trials of genetically engineered bacteria (called Ice Minus or Frostban) and, later, over genetically engineered foods. Now, California has the distinction of having the first and only law in the nation that bans human cloning.

Whether it is human cloning or genetically engineered food, all of the biotech controversies to date have shared a common thread. Put simplistically, scientific advances have been hyped by researchers, the press, and others as "revolutionary breakthroughs" in order to stir the imaginations (and open the pocketbooks) of investors, only to be seen by some other people as being so new and unfamiliar as to seem dangerous and threatening—the genie let out of the bottle or Pandora's box.

In each debate, the issues have consistently been framed in terms of public safety and the urgent need for immediate government intervention. I have learned two things from all of this. First, one person's revolutionary breakthrough may be another's *Andromeda Strain*. The absence of a common context to bridge these perceptions creates an unhealthy tension at the public policy interface that can seriously undermine sound decision making—and often does. Second, policies can be written in a timeframe of months, but can be corrected only over decades. Opportunities lost in the intervening time often are irretrievable.

Consider our history in biotech policymaking. Each major development in biotechnology—whether genetically engineered drugs or transgenic plants that need no pesticides—has been cast by proponents and opponents alike as fundamentally new and different (although for different reasons). The reports of Dolly's birth did the same thing.

For some people, new things are unfamiliar things. In policy debates, newness is often synonymous with riskiness, and risk, in the eyes of government, must be managed. Thus, the way biotechnology advances are being communicated to the public commonly casts new developments as potentially risky and warranting government attention.

In the scientific community, however, biotechnology developments are seen as incremental advances in a long continuum of life science research. They fit into a larger context of decades of basic research in biology, biochemistry, and genetics—and a related regulatory and legal structure. This view is adopted unevenly in federal policies.

For example in 1980, the Food and Drug Administration (FDA) considered whether the new drugs being produced using genetic engineering techniques were adequately overseen with existing regulations. They compared genetically engineered drugs with similar drugs produced using more conventional methods. They found no evidence of substantive differences in safety, efficacy, or quality and announced that there would be no new regulatory requirements.

In contrast, just a few years later, the U.S. Department of Agriculture (USDA) and the Environmental Protection Agency (EPA) announced that plants and microorganisms modified using genetic engineering techniques would be subject to special regulatory requirements. The new regulations turned on the use of genetic engineering techniques, which, in their view, were new and unfamiliar and hence potentially risky. The FDA considered the characteristic of the product and decided no new regulations were needed. The USDA and EPA focused on the technique and called for new regulations. The differential impact of these decisions on academic research and commerce has been substantial. For instance, the FDA applies the same regulatory requirements to insulin whether it is produced by genetic engineering, chemical extraction, or chemical synthesis.

The USDA applies different requirements to identical plants if one is produced using genetic engineering techniques and the other using traditional plant breeding methods. Only the genetically engineered plant is regulated.

The new regulatory requirements have increased the cost of early-stage field research and commercial development. They have discouraged investment in startup agricultural biotechnology companies by artificially raising market entry costs. Advances in agricultural biotechnology products have thus been effectively distanced from the marketplace and consumers. That is not to say that no important advances in agricultural biotechnology are reaching application. There are just many, many more that will not. Academic researchers and students are discouraged by the tremendous opportunities being lost.

In contrast, the FDA announcement of a level playing field for genetically engineered drugs sent a strong signal to the academic research community and fueled robust investment in the private sector in biopharmaceutical companies. The pace of delivery of academic advances in biomedical research to consumers has accelerated markedly.

Today, millions of Americans have benefited from biotechnology products, including clot-busting tissue plasminogen activator for heart attack victims and human insulin for diabetics, and our blood supply has been made safer with sensitive tests for the human immunodeficiency and hepatitis viruses.

The fundamental difference between the two policy approaches lies in the trigger for regulation: The FDA focuses on the characteristics of the end product (safety, efficacy, and quality), whereas the USDA and EPA focus on the technique or process (genetic engineering). As a result, the USDA and EPA regulate all genetically engineered products with the same high level of stringency because certain techniques were used.

The debate about the government's role in human cloning research is not unlike that of genetically engineered drugs and plants. There is a call for banning or regulating cloning, just as there has been for genetic engineering.

Cloning—whether we are talking about cloning DNA, cells, or whole organisms—is an integral part of the modern biotechnology toolbox. DNA molecules and individual cells are cloned every day in thousands of academic research laboratories around the world. The cloning of plants and animals is routine in commercial agriculture. *Cloning* has become a catchword for a variety of techniques used in research and commerce.

Cloning is not new. Current advances in mammalian cloning are part of a continually evolving string of research projects dating back to the turn of the century and the famous embryologist Hans Spemann. It also extends from decades of research in animal husbandry focused on improving livestock.

Recent advances in cloning methods applied to human cells, organelles, and DNA hold real potential in medicine. Somatic cell nuclear transfer (SCNT) technology is an important tool for studying embryonic development and the molecular and cellular events involved in birth defects, cancer, and aging.

It also holds promise for reducing or eliminating the risk of immune rejection in tissue transplantation by producing tissue in culture that is genetically identical to the patient or at least similar enough at the level of the cell surface to evade immune surveillance. Potential applications range from skin grafts for burn victims and bone marrow stem transplants for leukemia patients to nerve stem cell transplants for neurodegenerative diseases such as multiple sclerosis and amyotrophic lateral sclerosis.

We are just embarking on a new age of human gene therapy to treat disease. Cell cloning and SCNT technologies are considered potentially important strategies for enhancing the efficiency and efficacy of gene therapy.

A legislative ban or requirement for regulation of "cloning" or "human cloning" would potentially derail these important lines of research and many more. It is more than a problem in semantically refining the scope of the legislation's focus. Focusing on the tools at all runs the risk of taking them out of the hands of researchers who could put them to productive use for society.

Our experience with the policies of the USDA and EPA has shown that government restrictions on the technique or process can discourage research and commercial investment. The tools are made too expensive to use by virtue of the higher costs and paperwork associated with regulatory compliance. I often encounter academic researchers who are choosing to use cruder and less precise methods for making genetic modifications in plants simply because those older methods are not regulated.

The FDA avoided interfering in the researcher's choice of tools by focusing oversight on the safety, efficacy, and quality of the product. That enabled biomedical researchers in academia and industry to choose the best tool for the job, and research and innovation have flourished.

There is strong public demand for advances in medicine. I don't think anyone would argue that there is not also demand for genetically improved crop plants that require little or no chemical pesticides or fertilizers.

We are at a crossroads in medical science. We have a long tradition of strong public support for basic biomedical research. Since World War II, the federal government has supported broad and free exploration of the fundamental problems of health and disease. This has produced impressive advances in medicine.

The past two decades have fostered a true biological revolution that has produced powerful new research tools and unlocked the molecular basis of genes and heredity. The Human Genome Project is mining our genes for new ways to treat and prevent disease. The biotechnology industry is producing entirely new classes of drugs and vaccines that are tailored at the molecular level for greatest efficacy.

Rather than rushing to draw up legislation, we need thoughtful dialogue of the sort the University of San Francisco hosted in April 1998. Rather than restrict access to useful research tools, we need to describe more clearly the kinds of outcomes that are to be prevented. We then need to create an inventory of existing oversight mechanisms and determine whether they are inadequate before creating new laws or regulations.

For example, we have long used human subjects in biomedical and clinical research. The safety of human subjects in a broad spectrum of research projects is overseen by the National Institutes of Health (NIH; through the Office for Protection from Research Risks) at the level of the research proposal for NIH funding. This brings the proposed research into focus before the project is approved for funding and provides a mechanism for tracking the approved project over time.

The FDA also regulates the use of human subjects. Both the NIH and the FDA have stated that their safeguards extend to research involving cloning technologies.

Our track record provides a useful starting point for considering the adequacy of existing oversight mechanisms. Among other things, it draws attention to the oversight of in vitro fertilization clinics, where more effective control may be achieved by the development and certification of standards for clinic operation than by simply banning cloning tools.

In the spring of 1997 a number of national and international scientific societies, representing the majority of scientists who are knowledgeable in the science and tools necessary to accomplish cloning, called for a voluntary international moratorium on human nuclear transfer for the purpose of creating a new human being.

This calls to mind the voluntary moratorium scientists called for at the Asilomar conference in the 1970s. It sent a strong message that the scientific community understood the public concern and supported the need for dialogue before proceeding. But, as they say, no good deed goes unpunished.

The moratorium focused specifically on the use of recombinant DNA techniques, rather than on identifying the types of experimental organisms or genes that could pose serious risks and were to be avoided. Why does that matter? The Asilomar conferees called for the establishment of a Recombinant DNA Advisory Committee (known as the RAC) at the NIH and for formal guidelines and restrictions on the use of recombinant DNA techniques and molecules. The new restrictions were applicable to all research supported by the NIH.

At about this time, the construction of the new Molecular Biology Institute at the University of California at Los Angeles was nearing completion. The NIH rules required substantial new containment and confinement facilities at such great cost that the building project nearly stalled. The NIH RAC quickly found that the initial focus of the rules was too broad. Within two years of the Asilomar conference they exempted nearly 95 percent of all recombinant DNA research because they found that the vast majority was focused on benign organisms and genes that required no special precautions.

Over the next several years additional funds had to be raised to redesign the Molecular Biology Institute laboratories and return space devoted to special containment equipment back to productive research functions.

Perhaps more important, the Asilomar conference and the NIH RAC it established set a precedent for technique-based oversight. That precedent has been cited by the EPA and the USDA in defense of their regulatory approaches. It is also widely cited in Europe in support of stringent technique-based regulatory approaches and the nontariff trade barriers that they create.

The current moratorium on human cloning by scientific societies also focuses on technique: nuclear transfer. It runs the same potential risk of inadvertently reinforcing the view that government oversight should be focused on technique rather than outcomes and defined risks.

These are not simple issues. There are no simple solutions. We have learned that hastily drawn policies and calls for voluntary scientific moratoria can and do affect the course of research and eventual public benefits. When the public invests tax dollars in basic research and in government regulatory agencies, it reasonably expects thoughtful stewardship of those investments. We have nearly twenty years' experience in biotechnology policymaking. It can and should help us make the right decisions in the current debate over human cloning.

Selected Bibliography

SCIENCE

Briggs, R., and T. J. King. "Transplantation of Living Nuclei from Blastula Cells into Enucleated Frogs' Eggs." *Proceedings of the National Academy of Sciences, U.S.A.* 38 (1952): 455–63.

Campbell, K. H. S., J. McWhir, W. A. Ritchie, and I. Wilmut. "Sheep Cloned by Nuclear Transfer from a Cultured Cell Line." *Nature* 380 (1996): 64–66.

McKinnell, Robert G. *Cloning.* Minneapolis: University of Minnesota Press, 1985.

Schnieke, A. E., A. J. Kind, W. A. Ritchie, K. Mycock, A. R. Scott, M. Ritchie, I. Wilmut, A. Coleman, and K. H. Campbell. "Human Factor IX Transgenic Sheep Produced by Transfer of Nuclei from Transfected Fetal Fibroblasts." *Science* 278 (1997): 2130–33.

Seidel, F. "Die Entwicklungspotenzen Einer Isolierten Blastomere des Zweizelstadiums im Säugertierei [Developmental Potential of an Isolated Mammalian 2-Cell-Stage Blastomere]." *Naturwissenschaften* 39 (1952): 355–56.

Seidel, G. E., Jr. "Production of Genetically Identical Sets of Mammals: Cloning?" *Journal of Experimental Zoology* 228 (1983): 347–54.

———. "Superovulation and Embryo Transfer in Cattle." *Science* 211 (1998): 351–58.

Spemann, H. *Embryonic Development and Induction.* 1938. Rpt., New York: Hafner, 1967.

Willadsen, S. M. "A Method for Culture of Micromanipulated Sheep Embryos and Its Use to Produce Monozygotic Twins." *Nature* 277 (1979): 298–300.

Wilmut, I., A. E. Schnieke, J. McWhir, A. J. Kind, and K. H. S. Campbell. "Viable Offspring Derived from Fetal and Adult Mammalian Cells." *Nature* 385 (1997): 810–13.

ETHICS AND PUBLIC POLICY

Andrews, Lori B. *The Clone Age: Adventures in the New World of Reproductive Technology.* New York: Henry Holt, 1999.

Baltimore, David, Robert Coles, John B. Fagan, Immanuel Jakobovits, Bill Frist, Leon R. Kass, John O'Connor, Robert Siverberg, and Edward Teller. "Will Cloning Beget Disaster?" *Wall Street Journal,* May 2, 1997, p. A14.

Bonnicksen, Andrea L. "Ethical and Policy Issues in Human Embryo Twinning." *Cambridge Quarterly of Healthcare Ethics* 4 (Summer 1995): 268–84.

Brannigan, Michael C. *Ethical Issues in Human Cloning.* New York: Seven Bridges Press, 1999.

Cole-Turner, Ronald, ed. *Human Cloning: Religious Responses.* Louisville, Ky.: Westminster John Knox Press, 1997.

Eibert, Mark D. "Clone Wars." *Reason* 30:2 (June 1998): 52–54.

Fitzgerald, Kevin T. "Little Lamb, Who Made Thee?" *America,* Mar. 29, 1997, p. 3.

Fletcher, Joseph. *Ethics of Genetic Control: Ending Reproductive Roulette.* Buffalo, N.Y.: Prometheus, 1984.

Harris, Mark. "To Be or Not to Be." *Vegetarian Times* 250 (June 1998): 64–69.

Hefley, James C., and Lane P. Lester. *Can/Will Humans Be Cloned?* Ada, Mich.: Baker, 1998.

———. *Human Cloning: Playing God or Scientific Progress?* Ada, Mich.: Fleming H. Revel, 1998.

Hendin, Herbert. *Seduced by Death: Doctors, Patients, and the Dutch Cure.* New York: W. W. Norton, 1996.

Humber, James M., and Robert Almeder. *Human Cloning.* Totowa, N.J.: Humana, 1998.

Johnson, George. "Ethical Fears Aside, Science Plunges On." *New York Times,* Dec. 7, 1997, p. 6.

Kahn, Axel. "Clone Mammals . . . Clone Man?" *Nature* 386 (Mar. 13, 1997): 119.

Kass, Leon. "Wisdom of Repugnance." *New Republic,* June 2, 1997, pp. 17–26.

Kass, Leon, and James Q. Wilson. *The Ethics of Human Cloning.* Washington, D.C.: AEI Press, 1998.

Kilner, John F. "Stop Cloning Around." *Christianity Today* 41:5 (Apr. 28, 1997): 10–11.

Kimbrell, Andrew, and Bernard Nathanson. *The Human Body Shop: The Cloning, Engineering, and Marketing of Life.* 2d ed. Washington, D.C.: Regnery, 1998.

Kolata, Gina. *Clone: The Road to Dolly and the Path Ahead.* New York: William Morrow, 1998.

———. "On Cloning Humans, 'Never' Turns into 'Why Not.'" *New York Times,* Dec. 2, 1999, pp. A1, A24.

———. "With an Eye on the Public, Scientists Choose Their Words." *New York Times,* Jan. 6, 1998, p. F4.

Lester, Lane P., and James C. Hefley. *Human Cloning.* Ada, Mich.: Baker, 1998.

Lewontin, Richard. "Confusion over Cloning." *New York Review of Books*, Oct. 23, 1997, pp. 18–23.

Lewontin, Richard, Harold Shapiro, et al. "Confusion over Cloning: An Exchange." *New York Review of Books*, Mar. 5, 1998, pp. 46–47.

Macklin, Ruth, "Splitting Embryos on the Slippery Slope: Ethics and Public Policy," *Kennedy Institute of Ethics Journal* 4 (Sept. 1994): 209–25.

McCormick, Richard. "Blastomere Separation: Some Concerns." *Hastings Center Report* 24:2 (Mar./Apr. 1994): 14–16.

McGee, Glenn. *The Human Cloning Debate*. Berkeley, Calif.: Berkeley Hills Books, 1998.

———. *The Perfect Baby*. Lanham, Pa.: Rowman & Littlefield, 1997.

McGuen, Gary E., ed. *Cloning: Science and Society*. Hudson, Wis.: GEM Publishers, 1998.

Moraczewski, Albert S. "Cloning Testimony." *Ethics and Medics* 22 (May 1997): 3–4.

Morell, Virginia. "A Clone of One's Own." *Discover* 19:5 (May 1998): 82–90.

National Bioethics Advisory Commission (NBAC). *Cloning Human Beings*. Rockville, Md.: GEM Publishers, 1997.

Nussbaum, Martha, and Cass R. Sunstein, eds. *Clones and Clones: Facts and Fantasies about Human Cloning*. New York: W. W. Norton, 1998.

Pence, Gregory. *Flesh of My Flesh: The Ethics of Cloning—A Reader*. Lanham, Pa.: Rowman & Littlefield, 1998.

———. *Who's Afraid of Human Cloning?* Lanham, Pa.: Rowman & Littlefield, 1998.

Pojman, Louis P. *Life and Death*. Belmont, Calif.: Wadsworth, 2000.

Pontifical Academy for Life. "Human Cloning Is Immoral." *The Pope Speaks* 43 (Jan./Feb. 1998): 27–32.

Ramsey, Paul. *Fabricated Man*. New Haven, Conn.: Yale University Press, 1970.

Ramsey Colloquium. "On Human Rights." *First Things* 82 (Apr. 1998): 18–22.

Rantala, M. L., and Arthur J. Milgram, eds. *Cloning: For and Against*. Chicago: Open Court, 1999.

Rhodes, Rosamond. "Clones, Harms, and Rights." *Cambridge Quarterly of Healthcare Ethics* 4:3 (Summer 1995): 285–90.

Roberts, Melinda A. "Human Cloning: A Case of No Harm Done?" *Journal of Medicine and Philosophy* 21 (Oct. 21, 1996): 537–54.

Robertson, John A. "The Question of Human Cloning." *Hastings Center Report* 24:2 (Mar./Apr. 1994): 6–14.

Silver, Lee M. *Remaking Eden: Cloning and Beyond in a Brave New World*. New York: Avon, 1997.

———. *Remaking Eden: How Genetic Engineering and Cloning Will Transform the American Family*. New York: Avon, 1998.

"Uproar over Cloning." *Christian Century* 115:3 (Jan. 28, 1998): 76–77.

Walter, James J. "Theological Issues in Genetics." *Theological Studies* 60 (Mar. 1999): 124.

Winters, Paul A. *Cloning*. San Diego, Calif.: Greenhaven, 1998.

Contributors

ANDREA L. BONNICKSEN is professor and former chair of the Department of Political Science at Northern Illinois University, where she teaches courses in biomedical and biotechnology policy. She has written extensively on ethics and policy in genetics and health care, including *In Vitro Fertilization: From Laboratories to Legislatures* (1989). She is a member of the Ethics Committee of the American Society of Reproductive Medicine.

R. ALTA CHARO is an associate professor of law and medical ethics at the University of Wisconsin at Madison. She is on the faculties of the Medical School Program in Medical Ethics and the Law School, offering courses on health law, bioethics and biotechnology law, food and drug law, and medical ethics. The author of more than seventy-five articles, book chapters, and government reports, she has been a member of President Clinton's National Bioethics Advisory Commission since 1996.

JORGE L. A. GARCIA, a professor of philosophy at Rutgers University, has taught at Georgetown and Notre Dame universities, served as a senior research scholar at the Kennedy Institute of Ethics, and been a Fellow in Ethics at Harvard University. He is the author of numerous articles in theoretical and practical ethics, including contributions to *African-American Perspectives on Biomedical Ethics* (1992) and *"It Just Ain't Fair": The Ethics of Health Care for African-Americans* (1994).

SUSANNE L. HUTTNER directs the Systemwide Biotechnology Research and Education Program at the University of California (UC), which draws on researchers from the nine UC campuses, three UC-affiliated national laboratories, and the Agriculture Experiment Station to identify emerging opportunities and needs for biotechnology research and graduate training on biotechnology problems in the life sciences and chemical engineering, as well as supporting research in the social sciences and humanities on issues at the interface of biotechnology and society. As director of the BioSTAR Project at UC, she leads an industry-university partnership initiative that promotes technology transfer and supports research collaborations between UC scientists and California biotechnology companies.

PHILIP KITCHER is a professor of philosophy at Columbia University. The author of *Abusing Science* (1982), *The Nature of Mathematical Knowledge* (1983), *Vaulting Ambition* (1985), *The Advancement of Science* (1993), and *The Lives to Come* (1996), among other publications, he recently served as editor-in-chief of the journal *Philosophy of Science*. In 1993–94 he was a Senior Fellow at the Library of Congress, where he reported on the social implications of the Human Genome Project.

RICHARD C. LEWONTIN is the Alexander Agassiz Professor of Zoology at Harvard University and a professor of population sciences at the Harvard School of Public Health. An evolutionary geneticist who works on the genetic structure of natural populations, particularly at the molecular level, he is also active as a philosopher of science and has coauthored a number of papers on questions of evolutionary theory and its relationship to social issues. Among his published works are *Biology and Ideology* (1991), *Not in Our Genes* (1984), and *Human Diversity* (1982).

BARBARA MACKINNON is a professor of philosophy and chair of the Department of Philosophy at the University of San Francisco. Among her publications are *American Philosophy: An Historical Anthology* (1985) and *Ethics: Theory and Contemporary Issues* (2000, 3d ed.).

JOHN A. ROBERTSON holds the Vinson and Elkins Chair at the University of Texas School of Law at Austin. He has published widely on law and bioethics issues, including *The Rights of the Critically Ill* (1983), and written numerous articles on reproductive rights, organ transplantation, and human experimentation. Among his recent publications is *Children of Choice: Freedom and the New Reproductive Technologies* (1994).

GEORGE E. SEIDEL JR. is a professor of physiology and University Distinguished Professor at Colorado State University. A member of the National Academy of Sciences, he has received numerous awards for his work, including the Governor's Award for Science and Technology, Colorado (1986) and the Centennial Lectureship at Cornell University College of Veterinary Medicine (1994). His research has focused on reproductive manipulations in cattle and other agriculturally important animals, including horses, and he has written on various aspects of mammalian embryogenesis, fertilization, and cloning, including human cloning.

BONNIE STEINBOCK is a professor of philosophy in the Department of Philosophy at the University at Albany, State University of New York, with joint appointments in the Department of Public Policy at Rockefeller College and the Department of Health Policy and Management at the School of Public Health. Among her many publications are *Life before Birth: The Moral and Legal Status of Embryos and Fetuses* (1992), *Killing and Letting Die* (1994), and (with Dan Beauchamp) *New Ethics for the Public's Health* (1999).

Index

Typeset in 10.5/13.5 Adobe Minion
Composed by Celia Shapland
for the University of Illinois Press
Manufactured by Thomson-Shore, Inc.

University of Illinois Press
1325 South Oak Street
Champaign, IL 61820-6903
WWW.PRESS.UILLINOIS.EDU